Amphibian Biology

Edited by

Harold Heatwole
and
John W. Wilkinson

Volume 11

Status of Conservation and Decline of Amphibians: Eastern Hemisphere

Part 5

NORTHERN EUROPE

Published by Pelagic Publishing
www.pelagicpublishing.com
PO Box 874, Exeter, EX3 9BR, UK

Amphibian Biology, Volume 11: Status of Conservation and Decline of Amphibians: Eastern Hemisphere, Part 5: Northern Europe

ISBN 978-1-78427-016-2 (Pbk)
ISBN 978-1-78427-017-9 (ePub)
ISBN 978-1-78427-018-6 (ePDF)

British Library Cataloguing in Publication Data.

A catalogue record for this book is available from the British Library.

Cover image: The natterjack toad *Epidalea calamita* is a European endemic. The image shows a calling male in the city of Jena, Thuringia, Germany. In highly industrialized countries with a dense human population, the species often colonizes secondary habitats like quarry pits or wasteland within large cities. Photo: Andreas & Christel Nöllert.

Table of contents of volume 11, Amphibian Biology: Eastern Hemisphere, Part 5 (Northern Europe)

Contents of previous parts of volume 11, Amphibian Biology: Eastern Hemisphere

Part 2. North Africa (edited by Stephen D. Busack and Harold Heatwole) 2013, Basic and Applied Herpetology, Asociación Herpetológica Española, Madrid

Contributors to Part 5 (Northern Europe)

EDITORS

Heatwole, Harold, Department of Biology, North Carolina State University, Raleigh, NC 27695–7617, U.S.A., and Department of Zoology, The University of New England, Armidale, NSW 2351, Australia
harold_heatwole@ncsu.edu
Wilkinson, John W., Amphibian and Reptile Conservation, 655A Christchurch Road, Boscombe, Bournemouth, BH1 4AP, Dorset, U.K.
Johnw.wilkinson@arc-trust.org

AUTHORS

Adrados, Lars Christian, Årup Byvej 44, 7752 Snedsted, Denmark
lca@amphi.dk
Andersen, Andeas, Midtkobbel 73, 6440 Augustenborg, Denmark
A-Andersen@mail.dk
Andrén, Claes, Nordens Ark, Åby säteri, SE 456 93 Hunnebostrand, Sweden
claes.andren@artbevarande.se
Baláž, Vojtech, Department of Ecology and Diseases of Game, Fish, and Bees, Faculty of Veterinary Hygiene and Ecology, University of Veterinary and Pharmaceutical Sciences Brno, Palackého tř. 1946/1, CZ-612 42 Brno, Czech Republic
balazv@vfu.cz
Briggs, Lars, Amphi-consult, Forskerparken 10, 5230 Odense M, Denmark
lb@amphi.dk
Ceirans, Andris, Department of Zoology and Animal Ecology, Faculty of Biology, University of Latvia, Kronvalda bulvaris 4, LV-1010 Riga, Latvia
andris.ceirans@lu.lv
Christensen, Per Klit, Vistelhøjvej 5, 6933 Kibæk, Denmark
pkc@amphi.dk
Damm, Niels, Dalby Bygade 26, 5380 Dalby, Denmark
nd@amphi.dk
Engel, Edmée, Musée national d'histoire naturelle, 1A, rue Plaetis, L-2338 Luxembourg, Luxembourg
Edmee.ENGEL@mnhn.lu
Fog, Kåre, Hesselhom 107, 3670 Veksø, Denmark
kf@amphi.dk
Gjerde, Leif Yngve, Norske Naturveiledere (Naturopa Consultancies), Postboks 247, N-2001 Lillestrøm, Norway
post@naturveilederne.no
Griffiths, Richard A., Musée national d'histoire naturelle, 1A, rue Plaetis, L-2338 Luxembourg, Luxembourg
R.A.Griffiths@kent.ac.uk
Hansen, Finn, Birkevej 3, Nylars, 3720 Åkirkeby, Denmark
loevfroe@gmail.com

Hesselsøe, Martin, Amphi-Consult, Forskerparken NOVI, Niels Jernes Vej 10, 9220 Ålborg Øst, Denmark
mh@amphi.dk
Jeřábková, Lenka, Nature Conservation Agency of the Czech Republic, Kaplanova 1931/1, CZ-148 00, Praha 11 – Chodov, Czech Republic
lenka.jerabkova@nature.cz
Kautman, Ján, Slovak National Museum – Natural History Museum, Vajanského nábrežie 2, 810 06 Bratislava, Slovakia
jan.kautman@snm.sk
Mikkelsen, Uffe, Kongevej 48, 6100 Haderslev, Denmark
salamander@webspeed.dk
Mikulíček, Peter, Department of Zoology, Faculty of Natural Sciences, Comenius University in Bratislava, 842 15 Bratislava, Slovakia
pmikulicek@fns.uniba.sk
Nöllert, Andreas, Mönchsgasse 10, D-07743 Jena-Löbstedt, Germany
andreas.noellert@googlemail.com
Ojielska, Maria, Department of Evolutionary Biology and Conservation of Vertebrates, Institute of Environmental Biology, University of Wrocław, ul. Sienkiewicza 21, 50–335 Wrocław, Poland
maria.ogielska@uwr.edu.pl
Pabijan, Maciej, Department of Comparative Anatomy, Institute of Zoology and Biomedical Research, Jagiellonian University, ul. Gronostajowa 9, 30–387 Kraków, Poland
maciej.pabijan@uj.edu.pl
Podloucky, Richard, Heisterkamp 17, D-30916 Isernhagen, Germany
richard.podloucky@gmx.de
Proess, Roland, Umweltplanungsbüro Ecotop, 13, rue des Fraises, L-7321 Steinsel, Luxembourg
Email: ecotop@pt.lu
Pupina, Agnese, Latgales Ecological Society. Ainavas, Kalkunes pagasts, Daugavpils novads, LV-5449 Latvia
minka2112@inbox.lv
Pupina, Aija, Latgales Zoo. Vienibas street 27, Daugavpils. LV-5400, Latvia
(Deceased)
Pupins, Mihails, Institute of Life Sciences ad Technology, Daugavpils University, Parades street 1a, Daugavpils, LV5400, Latvia
mihails.pupins@gmail.com
Rannap, Riinu, Institute of Ecology and Earth Sciences, University of Tartu, Vanemuise 46, 51014 Tartu, Estonia
riinu.rannap@ut.ee
Rimšaitė, Jolanta, Nature Research Centre, Akademijos 2, Vilnius, 08412 Lithuania
entlab@gmail.com
Šandera, Martin, HERPETA, Šípková 1866/12, CZ-142 00 Prague 4, Polabské muzeum, Palackého 68, CZ-290 55 Poděbrady, Czech Republic
m.sandera@seznam.cz
Saarikivi, Jarmo, P.O. Box 65, FI-00014, University of Helsinki, Finland
saarikiv@mappi.helsinki.fi
Schley, Laurent, Administration de la nature et des forêts, 81 avenue de la Gare, L-9233 Diekirch, Luxembourg
laurent.schley@anf.etat.lu

Schmidt, Benedikt R., Koordinationsstelle für Amphibien- und Reptilienschutz in der Schweiz (karch), Unimail, Info Fauna Karch, UniMail G. Bellevaux 51, 2000 Neuchâtel, Switzerland
Benedikt.Schmidt@unine.ch

Sztatecsny, Marc, Department of Integrative Zoology, University of Vienna, Althanstrasse 14, 1090 Vienna, Austria
marc.sztatecsny@univie.ac.at

Trakimas, Giedrius, Institute of Biosciences, Life Sciences Centre, Vilnius University, Saulėtekio 7, LT-10257 Vilnius, Lithuania
giedrius.trakimas@gf.vu.lt

Vuorio, Ville, School of Forest Sciences, University of Eastern Finland, P.O. Box 111, FI-80101 Joensuu, Finland
ville.vuorio@uef.fi

Wood, Laura, Durrell Institute of Conservation and Ecology, School of Anthropology and Conservation, Marlowe Building, University of Kent, Canterbury, CT2 7NR UK
laura_rw@yahoo.com

Zumbach, Silvia, Koordinationsstelle für Amphibien- und Reptilienschutz in der Schweiz (karch), Unimail, Info Fauna Karch, UniMail G. Bellevaux 51, 2000 Neuchâtel, Switzerland
Silvia.Zumbach@unine.ch

54 Status and conservation of amphibians in Luxembourg

Laura R. Wood, Edmée Engel, Richard A. Griffiths, Roland Proess and Laurent Schley

Abbreviations and acronyms used in the text and list of references

ANF	Administration de la nature et des forêts
IUCN	International Union for the Conservation of Nature
LC	Least Concern
MNHNL	Musée national d'histoire naturelle [Luxembourg]
PA	Protected Area
PNPN	Plan national concernant la protection de la nature
SAC	Special Area of Conservation
SAP	Species Action Plan
SICONA	Syndicat intercommunal de la conservation de la nature.

I. Introduction

A. Luxembourg's geography

Situated between 49° and 51° northern latitude, and between 5° and 7° eastern longitude, Luxembourg is a small country in the heart of Western Europe, landlocked between Belgium, France and Germany. Despite its small size (2586 km²), it has diverse geology and landscapes. The northern part of the country, called 'Oesling', is part of the Ardennes Mountains that reach from France through Belgium, Luxembourg and into Germany, and is dominated by hills up to 560 m in elevation and steep valleys with small streams. Human population density is lower than the rest of the country. The southern part of the country, called 'Gutland', has an average elevation of 200 m and shows more gentle slopes and broader valleys than the Oesling. Whereas in the north, soils are rather infertile, the southern part is now largely dedicated to mixed agriculture.

The largest rivers of Luxembourg are the Moselle, which forms the border with Germany for some 30 km, the Sûre which cuts through northern Luxembourg from west to east to join the Moselle in the small town of Wasserbillig, and the Our, forming the border with Germany in the northeastern part of the country. Nearly all of Luxembourg's rivers and streams form part of the Moselle catchment.

Luxembourg's annual rainfall averaged 898 mm per year between 1981 and 2010, ranging from 58 mm to 87 mm per month (MeteoLux 2018); however, the total rainfall in 2017 was only 726 mm, and the average annual rainfall from 2012–2017 was 832.5 mm. The wettest months tend to be May–June and November–December, with the early summer rain helping to delay the drying of amphibians' breeding ponds, although a trend towards lower annual rainfall is concerning. Average monthly air temperatures in Luxembourg range from 0.8°C in January to 18.2°C in July (long-term average temperatures 1981–2010, MeteoLux 2018). In 2017, January was colder than average (–1.6°C) and July was hotter (19.1°C), but 2017 saw unprecedented summer temperatures for the country (MeteoLux 2018).

With a human population of nearly 600,000 (Anonymous 2015) there is ever growing pressure on natural habitats and wildlife in Luxembourg. According to national assessments, approximately 55% of Luxembourg's mammals, 40% of birds, 30% of reptiles, and 70% of amphibians are threatened (Wolff 2009). The key threats to wildlife in Luxembourg are considered to be intensive agriculture, forestry monocultures, urban development, pollution and fragmentation of the landscape (Wolff 2009).

B. Luxembourg's amphibians

There are 14 native species of amphibians known to be present in Luxembourg. Thirteen of these usually breed in ponds (*Triturus cristatus, Ichthyosaura alpestris, Lissotriton helveticus, L. vulgaris, Alytes obstetricans, Bombina variegata, Bufo bufo, Epidalea calamita, Hyla arborea, Pelodytes punctatus, Rana temporaria, Pelophylax lessonae*, including the hybrid species *Pelophylax* kl. *esculentus*), while *Salamandra salamandra* favours streams (Kwet 2009). One of these, *Pelodytes punctatus*, has only recently been found in the very south of Luxembourg, having previously been known from just across the border in France (Proess and Ulmerich 2013).

One of the major costs of the country's economic success has been the loss of biodiversity during the progression from a largely rural economy to a major European financial centre, via steel production (Chamber of Commerce of the Grand Duchy of Luxembourg 2009). The booming economy precipitated a rapid expansion of housing and business developments and also a vastly expanded transport network (Wolff 2009), particularly in the centre and south. However, despite this encroachment on natural habitats and one of the highest human population densities in Europe (Gaston *et al.* 2008), Luxembourg maintains a high percentage of forest cover (37.0%, mostly in the north) and a substantial coverage of protected areas of 20.2%, including Natura 2000 zones. The forested areas generally have a better conservation status than do farmed and aquatic ecosystems (Wolff 2009).

II. Conservation status and measures

Under the Bern Convention, all amphibian species occurring in Luxembourg are protected from exploitation or trade that could endanger whole populations; under Appendix 2 of the Convention, five species also have special protection against any deliberate disturbance (Table 54.1, based on Anonymous 1979). Furthermore, Annex IV of the Habitats Directive affords 'strict protection' to seven species, while *Triturus cristatus* and *Bombina variegata* are also protected under Annex II (Anonymous 1992), which stipulates that the 'core' areas of habitat should be protected under the Natura 2000 network.

Like 42% of amphibian species in the world (IUCN 2015), most of the species occurring in Luxembourg currently have a downward global population trend: 10 out of 14 species (including a hybrid) have been assessed to be decreasing across their global range, while the other four species are stable (Table 54.1). Despite the negative population trends, the IUCN global Red List

Table 54.1 Luxembourgish amphibians on the Bern Convention and Habitats Directive appendices and their IUCN Red List status and global population trend (LC = 'Least Concern').

Species	Bern Convention appendices[1]	Habitats Directive annexes[2]	Red List status[3]	Global population Trend[3]
Triturus cristatus	II	II, IV	LC	decreasing
Ichthyosaura alpestris	III		LC	decreasing
Lissotriton helveticus	III		LC	stable
Lissotriton vulgaris	III		LC	stable
Alytes obstetricans	II	IV	LC	decreasing
Bombina variegata	II	II, IV	LC	decreasing
Bufo bufo	III		LC	stable
Epidalea calamita	II	IV	LC	decreasing
Hyla arborea	II	IV	LC	decreasing
Pelodytes punctatus	III		LC	decreasing
Rana temporaria	III	V	LC	stable
Pelophylax lessonae	III	IV	LC	decreasing
Pelophylax kl. esculentus	III	V	LC	decreasing
Salamandra salamandra	III		LC	decreasing

[1] Anonymous 1979; [2] Anonymous 1992; [3] IUCN 2009.

categorised all amphibian species present in Luxembourg as 'Least Concern' in its 2008 assessment. This implies that they are considered to be widespread and/or abundant, but not that no action is needed: indeed, *'a taxon may require conservation action even if it is not listed as threatened'* (IUCN 2001).

As well as European protection laws, Luxembourg's national legislation affords protection to all amphibian species under the 2004 nature protection law (Anonymous 2004), detailed by the 2009 species-protection regulation (Anonymous 2009b). The drive for coherence and connectivity of protected areas (PAs) within and between EU countries (Haslett *et al.* 2008; Haslett *et al.* 2010; Samways *et al.* 2010) was pre-empted by several decades by an agreement between Belgium, the Netherlands and Luxembourg, formally called 'The Benelux Agreement concerning Nature Conservation and Landscape Protection' (Benelux Consultative Interparliamentary Council 1982). Such coordination of conservation efforts between small countries may help to improve efficiency by working within the context of a single larger area.

Monitoring and management of acutely threatened species and habitats in Luxembourg are implemented via national and regional governmental agencies or consultants directly contracted by the government. This ensures that adequate data are available to inform management decisions at specific locations. Biological records collected by volunteers and by people working under licence with protected species are collated by the Natural History Museum (Musée national d'histoire naturelle [Luxembourg] – MNHNL). These data are an important source of information for recording populations' locations and trends.

Luxembourg's own conservation targets, set out under the second edition of the 'Plan national concernant la protection de la nature' (PNPN), echo those of the Habitats Directive and the EU Convention on Biological Diversity. Listed as strategic priorities, these include: updating and implementing Natura 2000's plans for managing sites, implementing conservation measures, restoring degraded ecosystems, improving awareness and knowledge, and regular monitoring (Anonymous 2014). Since the first PNPN (Anonymous 2007), significant progress has been made on a number of targets, such as designation of more PAs and improving the habitats of threatened species (Anonymous 2009a, 2012).

According to national assessments by IUCN criteria, four amphibian species are considered at high risk within Luxembourg and each has its own Species Action Plan (SAP). *Triturus cristatus* is categorised as threatened (Groupe herpétologique du MNHNL 2009a), *H. arborea* is highly threatened (Groupe herpétologique du MNHNL 2009b), and *E. calamita* is at risk of extirpation (Proess 2009). The most recent addition to the SAP list is *A. obstetricans*, which is thought to be in severe decline, although it is currently classified as 'Least Concern' on Luxembourg's own Red List (Proess 2013). Work is underway to improve management of sites occupied by the SAP species and to create new breeding sites. A reintroduction programme for *H. arborea* that started in 2010 is so far judged to be a success.

One of the major benefits of the country's small size is that the conservation community is relatively small and everyone involved has a good knowledge of sites throughout the whole country. Information can be shared easily and tasks can be carried out promptly; however, communication between conservation professionals could still be improved. The government department for conservation (Administration de la nature et des forêts) (ANF) centrally coordinates conservation work.

III. Distribution of amphibians in Luxembourg

A. Protected areas

Protected areas (PAs) make up a significant part of many amphibian species' ranges in Luxembourg, serving an important role in their conservation. The five most common amphibian taxa in Luxembourg (*R. temporaria*, *Pelophylax* spp., *B. bufo*, *I. alpestris* and *L. helveticus*) and the slightly less common *S. salamandra* are found both inside and outside PAs, with distributions influenced by landscape. The rarest species, *H. arborea*, *B. variegata* and *E. calamita*, are restricted to 2–3 sites, not all of which are located within PAs, while *T. cristatus* and *L. vulgaris* do occur in PAs but are more common outside (Wood 2011).

B. Sizes of ranges

Although the known geographic range of *A. obstetricans* seems to have doubled over recent decades, this may well be due to the fact that this species has received considerable attention from conservationists, with more intensive searching being carried out (e.g. Wood 2011). Indeed, while its range has increased, numbers of individuals are thought to have plummeted, probably due to competition from other amphibians, and to agricultural intensification (Proess 2013), with no more than five calling males detected at most sites during recent surveys. The sizes of ranges of the common amphibian species remained stable or slightly increased between records collected pre-1985 and between 1996 and 2005. The picture is not so optimistic for other species, with the sizes of the ranges of *H. arborea*, *B. variegata*, *T. cristatus* and *L. vulgaris* all shrinking over the same period. The declines of *H. arborea* and *B. variegata* are especially alarming, with losses of 55.6% and 66.7% of former sites respectively between records collected pre-1985 and between 1996 and 2005 (Wood 2011). Both species were down to a single site each in 2010; for *B. variegata* a new site was discovered in 2012, and the aforementioned reintroduction project for *H. arborea* has increased the range of this species again; however, no further gains or losses have been reported since.

C. Species of conservation concern

Amphibians in Luxembourg are subject to threats similar to those faced by many amphibian species worldwide, including climatic change and modification and loss of habitat.

Chytridiomycosis (the disease caused by the fungus *Batrachochytrium dentrobatidis*) (*Bd*)) has also been detected in several locations, but the full extent of its impact locally is not known (Wood *et al.* 2009). A study on the impact of *Bd* on *A. obstetricans* found zoospore loads far below the threshold of 10,000 zoospores per swab determined for populations with epizootic outbreaks of chytridiomycosis, and concluded therefore that the infection with *Bd* was unlikely to be a major cause of the decline of this species in Luxembourg (Proess *et al.* 2015).

The distributions of the rarest species are shown in Figure 54.1. Historically, *E. calamita* was probably present throughout the country (Junck *et al.* 2003). However, since records began to be kept in 1955, it has only been recorded in five different locations, and most recently is only known from three sites. The remaining populations are geographically and genetically isolated from each other (Frantz *et al.* 2009). The sites currently hosting *E. calamita* require intensive management to prevent natural succession to forest, and suitable primary habitat is very limited within the country, whereas secondary habitat is also decreasing with the reduction of open quarrying.

In the five years up to 2000, the MNHNL database showed diminishing numbers of *H. arborea* at several sites in central Luxembourg, a more stable population at one south-eastern site, and a

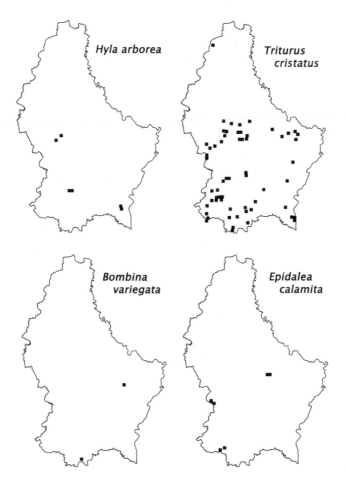

Fig. 54.1 The distributions of four threatened amphibian species in Luxembourg (1-km squares), 2003–2015. Data from the MNHNL database.

decline that reflects a trend across all of Western Europe (Stumpel 1997). Between 2000 and 2008 the species is thought to have been extirpated at all of the central sites (Wood 2011), leaving a single breeding pond in the south-east, which supports a large population and is surrounded by statutory protected areas. A habitat-creation and reintroduction project started at several sites in 2011 appears to be successful so far (Fig. 54.1).

There are thought to be two populations of *B. variegata* extant in Luxembourg, with one of the sites having been regularly documented since the early 1970s (Fig. 54.1). The origins of this population are subject to speculation about an unofficial reintroduction by a local person; however, these toads might just as easily have crossed the border from France without human assistance. Historically, Luxembourg was on the north-western edge of the species' European distribution (AmphibiaWeb 2015), but low rainfall and rising temperatures have shrunken the size of its range (Gollmann *et al.* 1997), perhaps making a wider distribution in Luxembourg unlikely. To improve the species' chances in Luxembourg, the large 'Special Area of Conservation' (SAC) that they inhabit has been managed to increase the number of shallow bodies of water, to boost its chances of successful breeding.

Triturus cristatus is relatively widespread compared to the other species detailed above (Fig. 54.1), but under ideal conditions it would be far more common. Its action plan (Groupe herpétologique du MNHNL 2009b) outlined a number of measures to be taken to reverse its decline, including the restoration of existing breeding sites, creation of more bodies of water to connect existing populations, extension of the protected areas surrounding breeding ponds, and development of suitable crossing tunnels for animals to use in migrating across roads or railways.

IV. Conclusions

Twenty percent of Luxembourg's territory has some form of statutory protection. Amphibians are relatively well monitored by professionals and volunteers, and data are collated centrally by the MNHNL. Most of the amphibian species present (73%) are assessed to be globally declining, with four of these meriting specific local action plans. It is clear that species' distributions and population dynamics are strongly influenced by those in neighbouring countries and the broader effect of global climatic change shifting or contracting species' ranges. However, despite the fact that funding for conservation in Luxembourg is still insufficient, there is a high concentration of experts on amphibians within this country and good collaborative relationships with the neighbouring countries exist, having led to several successful amphibian conservation projects in the recent past.

V. Acknowledgements

We thank all the scientific collaborators of the MNHNL that contributed to the collection of data. Most recent data resulted from the Natura 2000 monitoring programme set up by the Ministry of Sustainable Development and Infrastructures and carried out by the Luxembourg Institute for Science and Technology. SICONA also contributed a large quantity of recent data. Final thanks go to Donato Sereno (ANF) for help with GIS.

VI. References

AmphibiaWeb, 2015. *Information on amphibian biology and conservation* [web application]: Bombina variegata: *Yellow-Bellied Toad*. Berkeley, California: AmphibiaWeb. Electronic database available at: http://www.amphibiaweb.org/cgi/amphib_query?where-genus=Bombina&where-species=variegata (accessed on 6 August 2015).

Anonymous, 1979. Convention on the Conservation of European Wildlife and Natural Habitats. *Council of Europe Treaty Series* **104**: 1–12.

Anonymous, 1992. Council Directive 92/43/EEC of 21 May 1992 on the conservation of natural habitats and of wild fauna and flora. *Official Journal of the European Communities* **L206**: 7–49.

Anonymous, 2004. Loi du 19 janvier 2004 concernant la protection de la nature et des ressources naturelles; modifiant la loi modifiée du 12 juin 1937 concernant l'aménagement des villes et autres agglomérations importantes; complétant la loi modifiée du 31 mai 1999 portant institution d'un fonds pour la protection de l'environnement. *Mémorial A* **10**: 148–69.

Anonymous, 2007. Décision du Gouvernement en Conseil du 11 mai 2007 relative au plan national concernant la protection de la nature et ayant trait à sa première partie intitulée Plan d'action national pour la protection de la nature. *Mémorial A* **111**: 2034–47.

Anonymous, 2009a. Rapport de l'Observatoire de l'Environnement naturel. Ministère du Développement durable et des Infrastructures, Département de l'Environnement, Luxembourg.

Anonymous, 2009b. Règlement grand-ducal du 9 janvier 2009 concernant la protection intégrale et partielle de certaines espèces animales de la faune sauvage. *Mémorial A* **4**: 34–41.

Anonymous, 2012. Rapport de l'Observatoire de l'Environnement naturel. Ministère du Développement durable et des Infrastructures, Département de l'Environnement, Luxembourg.

Anonymous, 2014. Document de base en vue de la révision du Plan National concernant la Protection de la Nature. Ministère du Développement durable et des Infrastructures, Département de l'Environnement, Luxembourg.

Anonymous, 2015. *Luxembourg*. Electronic article available at: http://europa.eu/about-eu/countries/member-countries/luxembourg/index_en.htm (accessed on 16 June 2015).

Benelux Consultative Interparliamentary Council, 1982. *Benelux Convention on Nature Conservation and Landscape Protection*. Electronic article available at: http://sedac.ciesin.org/entri/texts/benelux.landscape.protection.1982.html (accessed on 4 August 2015).

Chamber of Commerce of the Grand Duchy of Luxembourg, 2009. *Your access to European markets: Luxembourg: where else?* Electronic article available at: http://eur-lex.europa.eu/LexUriServ/LexUriServ.do?uri=CELEX:31992L0043:EN:HTML (accessed on 5 August 2015).

Frantz, A.C., Proess, R., Burke, T. and Schley, L., 2009. A genetic assessment of the two remnant populations of the natterjack toad (*Bufo calamita*) in Luxembourg. *The Herpetological Journal* **19**: 53–9.

Gaston, K.J., Jackson, S.E., Nagy, A., Cantú-Salazar, L. and Johnson, M., 2008. Protected areas in Europe – principle and practice. *Annals of the New York Academy of Sciences* **1134**: 97–119.

Gollmann, G., Szymura, J.M., Arntzen, J.W. and Piálek, J., 1997. *Bombina variegata* (Linneaus, 1758). In: *Atlas of amphibians and reptiles in Europe*. Eds.: J.-P. Gasc, A. Cabela, J. Crnobrnja-Isailovic, D. Dolmen, K. Grossenbacher, P. Haffner, J. Lescure, H. Martens, J.P. Martínez Rica, H. Maurin, M.E. Oliviera, T.S. Sofianidou, M. Veith and A. Zuiderwijk, pp. 98–9. Societas Europaea Herpetologica and Muséum National d'Histoire Naturelle, Paris.

Groupe herpétologique du MNHNL, 2009a. *Plan d'action: Rainette arboricole (Hyla arborea)*. Plan national pour la protection de la nature (PNPN) – Plans d'actions espèces, Le Gouvernement du Grand-Duché de Luxembourg. Electronic article available at: http://environnement.public.lu/fr/natur/biodiversite/plan_d_action_especes_et_habitats/plan_d_action_especes.html (accessed on 11 June 2018).

Groupe herpétologique du MNHNL, 2009b. *Plan d'action: Triton crêté (Triturus cristatus)*. Plan national pour la protection de la nature (PNPN) – Plans d'actions espèces, Le Gouvernement du Grand-Duché de Luxembourg. Electronic article available at: http://environnement.public.lu/fr/natur/biodiversite/plan_d_action_especes_et_habitats/plan_d_action_especes.html (accessed on 11 June 2018).

Haslett, J.R., Berry, P.M. and Zobel, M., 2008. *European habitat management strategies for conservation: current regulations and practices with reference to dynamic ecosystems and ecosystem service provision*. The RUBICODE Project – Rationalising Biodiversity Conservation in Dynamic Ecosystems. Electronic article available at: www.rubicode.net/rubicode/RUBICODE_Review_on_Habitat_Management.pdf (accessed on 5 August 2015).

Haslett, J.R., Berry, P.M., Bela, G., Jongman, R.H.G., Pataki, G., Samways, M.J. and Zobel, M., 2010. Changing conservation strategies in Europe: a framework integrating ecosystem services and dynamics. *Biodiversity and Conservation* **19**: 2963–77.

IUCN, 2001. *IUCN Red List categories and criteria: version 3.1*. IUCN Species Survival Commission. IUCN, Gland and Cambridge.

IUCN, 2015. *The IUCN Red List of Threatened Species. Version 2015.2*. Available at: www.iucnredlist.org (accessed on 8 July 2015).

IUCN, 2017. *The IUCN Red List of Threatened Species. Version 2017–3*. Available at: www.iucnredlist.org (accessed on 11 June 2018).

Junck, C., Schoos, F. and Proess, R., 2003. *Bufo calamita*, Laurenti, 1768. In: *Ferrantia 37: Verbreitungsatlas der Amphibien des Großherzogtums Luxemburg*. Ed.: R. Proess. Musée National d'Histoire Naturelle, Luxembourg.

Kwet, A., 2009 *European Reptile and Amphibian Guide*. New Holland, London.

MeteoLux, 2015. Information on the climate in Luxembourg in 2014; Station: Luxembourg/Findel-Airport (WMO 06590, 376 m, a.s.l.); Reference period: WMO normal period 1961 to 1990. Available at: www.statistiques.public.lu/en/news/territory/territory-climate/2018/01/20180124/20180124.pdf (accessed on 26 June 2018).

Proess, R., 2009. *Plan d'action: Crapaud calamite (Bufo calamita)*. Plan national pour la protection de la nature (PNPN) – Plans d'actions espèces, Le Gouvernement du Grand-Duché de Luxembourg. Electronic article available at: http://environnement.public.lu/fr/natur/biodiversite/plan_d_action_especes_et_habitats/plan_d_action_especes.html (accessed on 11 June 2018).

Proess, R., 2013. Plan d'action Geburtshelferkröte – Alyte accoucheur, *Alytes obstetricans*. Electronic article available at: www.environnement.public.lu/conserv_nature/dossiers/Plans_d_actions/Plans_d_actions/PA_amphibiens_alytes_obstetricans.pdf (accessed on 15 July 2015).

Proess, R., Ohst, T., Plötner, J. and Engel, E. 2015 Untersuchungen zum Vorkommen der Geburtshelferkröte (*Alytes obstetricans*) und zur Verbreitung des Chytrid-Pilzes (*Batrachochytrium dendrobatidis*) in Luxemburg. *Bulletin de la Société des naturalistes luxembourgeois* **117**: 63–76.

Proess, R. and Ulmerich, M., 2013. Nachweis des Westlichen Schlammtauchers (*Pelodytes punctatus* (Daudin, 1802)) im Süden Luxemburgs (Amphibia, Anura, Pelodytidae). *Bulletin de la Société des naturalistes luxembourgeois* **114**: 89–92.

Samways, M.J., Bazelet, C.S. and Pryke, J.S., 2010. Provision of ecosystem services by large scale corridors and ecological networks. *Biodiversity and Conservation* **19**: 2949–62.

Stumpel, A., 1997. *Hyla arborea* (Linneaus, 1758). In: *Atlas of amphibians and reptiles in Europe*. Eds.: J.-P. Gasc, A. Cabela, J. Crnobrnja-Isailovic, D. Dolmen, K. Grossenbacher, P. Haffner, J. Lescure, H. Martens, Martínez Rica, J.P., H. Maurin, M.E. Oliviera, T.S. Sofianidou, M. Veith and A. Zuiderwijk, pp. 124–5. Societas Europaea Herpetologica and Muséum National d'Histoire Naturelle, Paris.

Wolff, F., 2009. Quatrième rapport national de la Convention de la diversité biologique du Grand-Duché de Luxembourg. Ministère du Développement durable et des Infrastructures, Luxembourg. Electronic article available at: www.cbd.int/doc/world/lu/lu-nr-04-fr.pdf (accessed on 5 August 2015).

Wood, L.R., 2011. Diversity and Distribution of Amphibians in Luxembourg. PhD thesis. University of Kent.

Wood, L.R., Griffiths, R.A. and Schley, L., 2009. Amphibian chytridiomycosis in Luxembourg. *Bulletin de la Société des naturalistes luxembourgeois* **110**: 109–14.

55 Status and conservation of amphibian species in the Federal Republic of Germany

Richard Podloucky and Andreas Nöllert

Abbreviations and acronyms used in the text and references

ABS	*Amphibien-Reptilien-Biotopschutz Baden Württemberg*
AGAR	*Arbeitsgemeinschaft Amphibienschutz und Reptilienschutz in Hessen*
Bd	Batrachochytrium dendrobatidis
BfN	*Bundesamt für Naturschutz*
BLAK	*Bund-Länder-Arbeitskreis*
BUND	*Bund für Umwelt und Naturschutz Deutschland*
CR	*Rank of "Critically Endangered" according to the Red List of the IUCN*
DD	*Rank of "Data Deficient" according to the Red List of the IUCN*
DGHT	*German Society for Herpetology and Herpetoculture*
DNA	*Deoxyribonucleic acid*
EN	*Rank of "Endangered" according to the Red List of the IUCN*
EU-LIFE	*European Union – L'Instrument Financier pour l'Environnement*
FFH-Monitoring	*Fauna-Flora-Habitat-Monitoring*
IUCN	*International Union for the Conservation of Nature*
LARS	*Landesverband für Amphibien und ReptilienSchutz in Bayern*
LC	*Rank of "Least Concern" according to the Red List of the IUCN*
NABU	*Naturschutzbund Deutschland*
NRW	*North Rhine Westphalia*
NT	*Rank of "Near Threatened" according to the Red List of the IUCN*
VU	*Rank of "Vulnerable" according to the Red List of the IUCN*

I. Introduction

A. Climate

The Federal Republic of Germany is located in Central Europe, centered on 51.1° N and 10.4° E. It is bordered to the north by Denmark and the North and Baltic Seas; to the south by Switzerland

and Austria (Alpine region); to the east by Poland and the Czech Republic; and to the west by the Netherlands, Belgium, Luxembourg and France. Germany covers an area of approximately 357,000 km² and is populated by about 82.2 million inhabitants (230 inhabitants/km²). With the exception of the Wadden Sea and some alpine areas, the landscape is strongly anthropogenically influenced, intensively cultivated and urbanized, rendered impermeable by artificial substrates, and highly fragmented by roads, railways and ship canals; 51.6% of the state's area is used for agricultural purposes, 30.6% is covered by forests, 13.7% comprises settlements and roadways, 2.4% consists of surface water (rivers and lakes) and 1.7% is of miscellaneous other landscapes (including wasteland) (Statistisches Bundesamt 2017).

The largest area of natural landscape is comprised of the glacial North German coastal lowlands (with the lowest point at 3.54 m below sea level). To the south are the low-mountain region and the steppeland, the foothills of the Alps and a narrow strip of the alpine mountain range (with the Zugspitze as the highest mountain, at 2,962 m above sea level) (Bundesamt für Naturschutz 1995). Germany is situated in the Central European summer-green, deciduous-forest zone, where (natural) beech and mixed forests predominate. The dominant soils are boulder clay, calcareous glacial till, sand, limestone, marlstone and dolomite rocks, clay and siltstones, respectively rocks of magmatic and metamorphic origin.

The river systems with the largest catchment areas are the Rhine, Ems, Weser, Elbe (draining into the North Sea), Oder (Baltic Sea) and Danube (Black Sea). Natural lakes are concentrated in the north in Holstein Switzerland and the Mecklenburg Lake District, and in the south in the Alpine foothills. Natural ponds, mainly of glacial origin (Sölle), are found in the north (Ostholstein, Mecklenburg-Vorpommern, Brandenburg) (Pätzig 2010) and the alpine foothills. Moors (4% of the land area) of different nutrient composition and hydrological characteristics are concentrated in the north-eastern part of Germany (fens), the north-western lowlands (50% raised bogs and 50% fens) and the alpine foothills (30% raised bogs, 70% fens); only 5% are in a natural state.

Landscapes with large ponds occur in the Oberlausitzer Heidelandschaft and Teichlandschaft (Saxony) as well as in Franconian Bavaria (Aischgrund) and the Tirschenreuther Teichpfanne (Upper Palatinate) among other places; these areas had already been adapted for fish-farming by the 10th–11th century.

Germany is in the transitional area between the Atlantic (western) and the Continental (north-eastern) climatic zones. The mean annual temperature for the period 1981–2010 was 8.9°C; the average monthly temperatures are between –0.4°C (January) and 18.0°C (July). The average annual rainfall is 819 mm. There is a tendency in this region towards increasing elevation of summer temperatures as well as of earlier springs than previously (https://de.wikipedia.org/wiki/Klima_in_Deutschland).

B. The amphibian fauna

There are 20 indigenous amphibian species in the Federal Republic of Germany. The edible frog (*Pelophylax* kl. *esculentus*), here treated as a separate species, is in reality a hybridogenetic taxon between the pool frog (*Pelophylax lessonae*) and the marsh frog (*Pelophylax ridibundus*) (Schulte 2014). Also, part of the hybrid zone between *Triturus cristatus* and *Triturus carnifex* (Maletzky *et al.* 2008) is located in the district of Berchtesgadener Land (Bavaria).

In addition to native species, there are two alien species that have become established in Germany. These are an allochthonous, regularly reproducing population of *Triturus carnifex*, which was introduced from Croatia in 1990–1991 (Franzen *et al.* 2002), and the American bullfrog (*Lithobates catesbeianus*), which exists as a reproducing population since at least the year 2000 in the Oberrheinebene north of Karlsruhe. Despite measures to combat it, it shows a tendency to spread.

Most of the native amphibian species are widespread in Europe, so there are no German endemic ones. However, owing to the country's location in the centre of the semi-humid European climatic region with predominantly deciduous and mixed forests and numerous central mountains, many species have a distributional focus within Germany (e.g. the fire salamander *Salamandra salamandra*, alpine newt *Ichthyosaura alpestris* and yellow-bellied toad *Bombina variegata*). The same is true for species that are widespread in Europe but in highly isolated patches at the borders of their ranges, e.g. relict populations (so-called outposts), such as those of the agile frog (*Rana dalmatina*) or moor frog (*Rana arvalis*). This high level of importance of Germany for core populations also applies to other species such as the great crested newt (*Triturus cristatus*), natterjack toad (*Epidalea calamita*) and edible frog (Kühnel *et al.* 2009).

Some amphibian populations in Germany are at, or very near, the borders of their range (SEH 2014; Sillero *et al.* 2014; IUCN 2017):

- Northern border of the range: alpine salamander (*Salamandra atra*) (Baden-Wurttemberg and Bavaria between approximately 600 m and 2100 m above sea level).
- Northern, western and/or eastern borders of the range: *Salamandra salamandra* and the palmate newt (*Lissotriton helveticus*) (Lower Saxony/Elbe, northern, north-western, and north-eastern borders of their ranges); *Ichthyosaura alpestris* (north-western, north-eastern borders of its range); common midwife toad (*Alytes obstetricans*) (Lower Saxony/northern edge of the central mountains, northern and eastern borders of its range); *Bombina variegata* (Lower Saxony/northern edge of the Central mountains, north-western, northern, north-eastern borders of its range), *Rana dalmatina* (northern, north-western, north-eastern borders of its range, where there are strongly fragmented occurrences).
- Western border of the range: fire-bellied toad (*Bombina bombina*), green toad (*Bufotes viridis*)
- South-western border of the range: *Rana arvalis*

The protection of wild animals and plants was established early by the Reichsnaturschutzgesetz (1935) and the Naturschutzverordnung (1936). Accordingly, the fire and alpine salamanders, toads, fire-bellied and yellow-bellied toads, treefrog (*Hyla arborea*), and all frogs except the common frog (*Rana temporaria*) and water frogs, have been protected since 1936. According to the Basic Law of the Federal Republic of Germany, this protection continued as a state law until 1986. In 1986, the Federal Wildlife Conservation Ordinance, in conjunction with the Federal Nature Conservation Act, specifically protected all indigenous amphibian species. This means that these species cannot be caught, kept, traded, injured or killed. The developmental stages (eggs, larvae) are also protected. Furthermore, sites for reproduction (bodies of water) or for shelter (e.g. hibernation sites) may not be damaged or destroyed. Eleven of these species are classified as strictly protected through part of the implementation of the European Union's Habitats Directive. In addition to the prohibitions mentioned above, they may not be significantly disturbed. In the Habitats Directive, three species are listed in Annex II (species of community interest whose conservation requires the designation of special areas of conservation), 11 species are listed in Annex IV (species of community interest in need of strict protection) as well as three species in Annex V (species of community interest in the wild and exploitation may be subject to management measures). In addition, amphibian sites can also be protected by "Protected Landscape Elements" or *per se* by "Legally Protected Biotopes" (e.g. natural or near-natural watercourses and ponds).

The first Red List of the endangered amphibians of Germany was produced in 1976 (Blab and Nowak 1976). The current Red List was published by Kühnel *et al.* (2009). According to the former, eight (40%) of 20 amphibian species belong to some category of threat and two are Near Threatened. In contrast, in 1998 (Beutler *et al.* 1998), 13 (62%) of 21 amphibian species were listed in one of the threat categories and two as Near Threatened. At the former time, the alpine crested newt (*Triturus carnifex*) was still considered indigenous, but this is not true according to recent findings. However, the lower number of threatened species is not related to an improvement in the existing situation, but to application of more stringent and clearer objective criteria. Involved in the assessment, and therefore responsible for the classification of the category of threat, are the current status and the long-term and short-term trends of populations and the degree of risk imposed by various factors (see Kühnel *et al.* 2009). The regionalization of the Red List also shows that the proportion of threatened species in the North German Lowlands (nine out of 19 species = 47%) is lower than in the mountainous and hilly country (11 out of 20 species = 55%) and in the Alps (eight out of 13 species = 61.5%). For some species, there is an east-west gradient. Thus, the degree of threat for *Bombina bombina* and *Bufotes viridis* (species with an "eastern", continental distribution) increases from east to west, whereas for *Epidalea calamita* ("western" distribution) it decreases.

Since in Germany the responsibility for the conservation of nature falls within the sovereignty of the federal states, regionalized Red Lists occur for almost each state (see table in Kühnel *et al.* 2009).

C. Amphibian decline and its causes

The amphibian species native to Germany are today threatened by the loss, fragmentation and quality of habitats, as well as by the effects of intensified agriculture (e.g. large-scale land use; use of fertilizers, herbicides and pesticides; intensive animal husbandry; frequent crop rotation; large-scale production of maize for biogas plants) (Landesumweltamt Brandenburg 2004). In addition, frequent destruction and impairment of their natural habitats has led to serious declines of amphibians' populations:

- watercourses are regulated and canalized, thereby eliminating large areas of riverside forests and floodplains as well as damaging natural dynamics of rivers through maintenance of watercourses, such as frequent dredging and the mowing of banks;
- filling-in of small ponds; the loss in the past decades is 40–50%, and regionally even 80% or more;
- melioration and use of fens and raised bogs with consequent lowering of the water table (drainage);
- fragmentation and loss of habitats as a result of ever-increasing urbanization, and the dissection of the landscape by roads, canals, and railways;
- genetic isolation or depletion;
- predation by domestic animals, waterfowl and fish;
- introduction of fish into small ponds;
- introduction of invasive species, e.g., *Lithobates catesbeianus* with its concomitant predation on native frogs and other animals;
- infectious diseases;
- inputs of nitrogen leading to an increase in the rate of succession in bodies of water.

Secondary biotopes, such as military training sites, excavation pits or near-natural fish ponds, which through anthropogenic influence, result in similar structures and dynamics to the original habitats, are becoming of increasing importance.

Another possible cause of amphibian declines is climatic change, but that cannot be assessed at the present time.

The fungus *Batrachochytrium dendrobatidis* (*Bd*) was first detected in Germany in 2000, although population declines have not yet been documented (Ohst *et al.* 2013). However, Ohst *et al.* (2011) detected *Bd* DNA in approximately 34% of the 156 amphibian populations analyzed between 2003 and 2010 and in 7.5% of the 3,013 specimens tested. Apart from *Salamandra atra*, all indigenous amphibian species were included in the sampling. The highest prevalences were found in *Ichthyosaura alpestris*, *Bombina variegata*, and *Pelophylax* spp. Böll *et al.* (2014) found particularly high prevalences in hibernated *Alytes obstetricans* tadpoles in the Bavarian Rhön, but no mortality was detected. In addition, a nationwide research programme, instigated in 2015, intended to monitor the spread of the recently-discovered (in the Netherlands) *Batrachochytrium salamandrivorans*; this fungus particularly affects *Salamandra salamandra*, but also other caudate amphibians in the long term (Martel *et al.* 2013, Sabino-Pinto *et al.* 2015). In the Eifel (Rhineland-Palatinate, North Rhine Westphalia), fire salamanders affected by this fungus had already been detected by targeted monitoring. Further spread is expected (Spitzen-van der Sluijs *et al.* 2016).

II. Declining species and species of special conservation concern

Detailed presentations and distribution maps (topographic map 1:25.000 grid) for the following species are found in Günther (1996), currently at DGHT (2014) and in Schulte (2014). The red-list status for all species has been taken from the Red List of Amphibians in Germany (Kühnel *et al.* 2009).

A. *Salamandra salamandra*

1. Distribution: The fire salamander is represented in Germany by two subspecies (*S. s. salamandra*, *S. s. terrestris*). Their main distribution is in the low-mountain region. The limit of distribution ends in the north at the foot of the mountain range. In the north German lowlands, the species is mostly absent, except for a series of isolated occurrences north-west, south and west of the river Elbe. At the same time, they represent the northernmost occurrences.

2. Habitat: The fire salamander prefers damp, cool deciduous and mixed forests with small brooks. These habitats often have ground-level soils with a wide range of hiding-places, such as dead wood, tree roots, stumps, burrows of small mammals and clefts in rocks. Small, fish-free spring-brooks with still bays and scours, as well as ponds, tarns, ditches, lanes and mining galleries are sites used for the deposition of the larvae.

3. Status and decline: This species is moderately common in Germany, and is frequent in the mountains, but rare in the lowlands and the Alps; declines are regionally variable from light in some low and high mountains, to strong in the Alps.

4. Red List ranking in Germany: This species is ranked as of "Least Concern (LC)" for some mountainous areas, as "Near Threatened (NT)" in low areas, and as "Endangered (EN)" in the alpine region.

5. International importance of German populations: Germany contains a significant part of the distributional range of this species (>10%) and is centrally located within that range.

6. Protection status: This species is especially protected.

7. Conservation measures and programmes: The species' benefits accrue nowadays from ecological management of state forests, and from the European Water Framework Directive for the protection and preservation of watercourses.

B. *Salamandra atra*

1. Distribution: The alpine salamander reaches its northern distributional limit in the extreme south of Germany along the edge of the Alps. Its main distribution is in the Bavarian Alps. The Northern Limestone Alps are mainly populated from about 800–2,100 m above sea level.

2. Habitat: In southern Germany, the Alpine salamander inhabits cool deciduous and mixed forests in valleys, as well as herbaceous, damp, mixed forests in the mountains. It occurs in clearings, at the edges of forests and near brooks. However, it is also found in open terrestrial habitats such as scree slopes and in wet alpine pastures in the dwarf shrub region above the limit of forests.

Since this species gives birth to fully developed young salamanders, it is independent of bodies of water; however, it is dependent on high humidity.

3. Status: Owing to its very limited distributional area, this species is generally rare; although moderately frequent in the Alps, in the mountains outside the Alps it is extremely rare. Its populations are stable.

4. Red List ranking in Germany: This species' ranking is LC; it is rare in mountains other than the Alps.

5. Protection status: This species is especially and strictly protected. It is listed in Annex IV of the Habitats Directive of the European Union (EU).

C. *Lissotriton helveticus*

1. Distribution: The distribution of the palmate newt is essentially restricted to the low-mountain region of western and central Germany. In the north-western part of the German lowlands there are numerous isolated occurrences, whereas it is absent in the north-eastern lowlands and in south-eastern Germany. The palmate newt reaches its northern and eastern distributional limit in Germany.

2. Habitat: The main habitat of the palmate newt consists of contiguous deciduous and mixed forests, and occasionally natural coniferous forests in the low-mountain and hilly regions. It is also found in open landscapes with only a few scattered trees. It uses all types of standing and weakly flowing waters for spawning. Usually these are small, cool bodies of water, such as pools, water-filled lanes and upper reaches of brooks, as well as dams and large ponds in habitats ranging from full exposure to the sun to semi-shaded.

3. Status and decline: This species is moderately frequent in Germany, being frequent in the mountains, but very rare in the lowlands; its population trend is generally stable, but in the lowlands it is experiencing a long-term moderate decline.

4. Red List ranking in Germany: This species is ranked as of LC, except in the lowlands where it is considered to be "Vulnerable (VU)".

5. Protection status: This species is especially protected.

6. Conservation measures: New installation and restoration of ponds, and removal of fish.

D. *Triturus cristatus*

1. Distribution: The great crested newt is widespread in Germany and is typically found on the plains and in hilly regions. In the Atlantic-influenced Northwest it is rare in the geest (sandy heath) landscape and is missing in the marshes. Population densities in the low mountains are markedly lower than elsewhere; at many localities, the species is missing. The Alps are not populated except for some valleys.

2. Habitat: As an open-land dweller, the great crested newt populates grasslands and agricultural land of lowlands, hilly places with adjoining deciduous and mixed forests, as well as networking structures such as hedges and spinneys. Secondary habitats such as gravel and clay pits are also of great importance. The medium to large, often permanent, waters used for spawning are characterized by exposure to the sun and with abundant shore and aquatic vegetation, especially ditches in meadows, fish-free ponds and backwaters.

3. Status and decline: This species is frequent in Germany, but usually very rare in the alpine region. It suffers long-term moderate declines in the lowlands, strong decline in the mountainous areas, except in the Alps where decline is very severe.

4. Red List ranking in Germany: This species is ranked generally as NT, but as LC in the lowlands, NT in the mountains except for the Alps, where it is considered to be "Critically Endangered" (CR).

5. International importance of German populations: Germany contains a rather extensive part of the distributional range of this species (10–30% or more) and is centrally located within that range.

6. Protection status: This species is especially and strictly protected; it is listed in Annex IV of the EU's Habitats Directive.

7. Conservation measures: This species occurs in designated protected areas (Annex II of the European Union's Habitats Directive); new ponds have been established and others restored; fish have been removed from ponds.

E. *Alytes obstetricans*

1. Distribution: The common midwife toad populates the low mountains of western and middle Germany as well as peripheral regions. Within the country, it also reaches its northern and eastern distributional limit.

2. Habitat: After the spoiling of rivers and brooks in the low mountains and hilly region, the common midwife toad now depends on sun-lit secondary habitats such as quarries, gravel, sand, clay pits and industrial fallow areas. Many hiding places, such as burrows, embankments, scree slopes, stone heaps and broken stone walls are used. Tadpoles use various sizes of bodies of water, such as pools, ponds, sinkholes, montane lakes and slowly flowing brooks.

3. Status and decline: This species is generally rare and is extremely rare in the lowlands; it experiences moderate to heavy long-term declines.

4. Red List ranking in Germany: In general, this species is considered "Vulnerable (VU)", but in the lowlands it is "Critically Endangered (CR)".

5. Protection status: This species is especially and strictly protected and is listed in Annex IV of the European Union's Habitats Directive.

6. Conservation measures: There has been installation of new ponds and restoration of existing ones, as well as restoration of terrestrial habitat.

F. *Bombina bombina*

As a typical lowland species north of the low-mountain ranges, the fire-bellied toad inhabits only north-eastern and eastern Germany and reaches its western boundary in eastern Schleswig-Holstein (east coast to the Baltic Sea), as well as the Elbe valley east of Hamburg.

1. Habitat: The natural habitats of the fire-bellied toad include open, sun-lit floodplains of semi-natural rivers with wet grassland and reedbeds, as well as adjoining forest ridges, carr and river-side forests, and spinneys. It spawns in medium to large, permanent, vegetation-rich, sun-lit standing waters with pronounced shallow zones, e.g., ponds, oxbows, hollows with waterlogged soil in the agricultural landscape, as well as temporarily flooded areas (return seepage).

2. Status and decline: This species generally is rare but moderately frequent in the lowlands, and extremely rare in the low-mountain region. It experiences both short-term and long-term strong declines generally, and very strong declines in the low-mountain region.

3. Red List ranking in Germany: This species is listed as EN generally, as VU in the lowlands, and as CR in the mountains.

4. Protection status: This species is especially and strictly protected and is listed in Annex IV of the EU's Habitats Directive.

5. Conservation measures: This species occurs in designated protected areas (Annex II of the EU's Habitats Directive) and is included in regional conservation programmes, such as installation of new ponds and restoration of old ones on a large scale. It is the subject of residents' support and of reintroduction following captive breeding.

G. *Bombina variegata*

As a typical inhabitant of mountainous and hilly areas, the yellow-bellied toad colonized the southern and western parts of Germany. It reaches the north-western, northern and north-eastern boundaries of its range at the transition zone with the North German lowland, where it is largely absent. The existing populations are patchy and isolated from each other.

1. Habitat: Originally, the yellow-bellied toad lived on the natural floodplains of streams and sun-lit forests containing temporary source-pools, scour basins and game wallows. Today it has been almost completely pushed back into secondary habitats such as mining pits, military training grounds, or industrial fallow areas, in which certain dynamics are maintained through human activities or by grazing animals, and where small, shallow, temporary pools, containing little or no vegetation, are used for spawning.

2. Status and decline: This species occurs moderately frequently in Germany generally, but it is extremely rare in the lowlands and rare in the Alps. It has experienced short-term and long-term strong to very strong declines and in the populations in the Alps there is a constant trend of moderate decline.

3. Red List ranking in Germany: This species is considered EN generally, CR in the lowlands, and NT in the Alps.

4. International importance of German populations: Germany contains an extensive part of the distributional range of the nominate subspecies *B. v. variegata*) (33%) and is centrally located within that range.

5. Protection status: This species is especially and strictly protected and is listed in Annex IV of the EU's Habitats Directive.

6. Conservation measures: This species is partly shielded from danger by inhabiting internationally designated protected areas (Annex II of the European Union's Habitats Directive) and being subject to nationwide and regional conservation programmes, such as the installation of new ponds and restoration of existing ones on a large scale, receiving resident support, and being included in reintroduction projects following captive breeding.

H. *Pelobates fuscus*

1. Distribution: The common spadefoot has its distributional focus in the lowland and hilly regions of northern and eastern Germany but is rare in the extreme north-west. In the western and southern parts of Germany, it occurs only regionally, especially on the plain of the upper Rhine. Its southern limit of distribution is along the Danube Basin in Bavaria. It does not occur in the Alpine area.

2. Habitat: As primarily a dweller of steppes, the common spadefoot is dependent on open terrestrial habitats with dry, loose, friable soils. Today, it is primarily found on sandy or loamy soils in agricultural land that is not used too intensively; in catchment areas of larger rivers or in inland dunes; heathland; sand and gravel pits; and industrial fallow areas. Larger, eutrophic, vegetation-rich ponds; floodplains; and ditches in meadows and the margins of lakes serve as habitats for spawning.

3. Status and decline: Moderately frequent in Germany generally, frequent in the lowlands, rare in the low-mountain region; trend from moderate, long-term, strong declines to constant moderate declines in populations in the lowlands; in the low-mountain region declines are strong to very strong.

4. Red List ranking in Germany: This species is listed as VU in general, but as LC in the lowlands, and CR in montane regions.

5. Protection status: This species is especially and strictly protected and listed in Annex IV of the EU's Habitats Directive.

6. Conservation measures: This species has benefited from installation of new ponds and restoration of extant ones, local conservation programmes, and reintroductions following captive breeding.

I. *Bufotes viridis/Bufotes variabilis*

1. Distribution: In Germany, the green toad mostly inhabits the lowlands and hilly regions. Three centers of widespread distribution can be determined: (1) eastern Germany, (2) the western and south-western part of the country (e.g., Rhine Valley) and (3) southern Germany. The north-western and western boundaries of this species from the thermophilic steppe runs from the far north of Germany across the Rhine Valley to France. Owing to genetic differences, two species now are distinguished. In contrast to the above-mentioned distribution of *Bufotes viridis*, *B. variabilis* appears to inhabit only the northern part of Germany (Stöck *et al.* 2008). However, since its actual distribution is still unclear and the taxonomic status of *B. variabilis* remains controversial and its validity is not accepted generally, this taxon is not treated further here.

2. Habitat: The green toad is found mainly in sunny, exposed habitats with loose soils, and occasionally in sandy floodplains and "Börde" landscapes. Today, it inhabits secondary habitats such as sand and gravel pits, fallow/ruderal fields with poor vegetation, and oligotrophic grassland. For spawning, it requires small to medium-sized, vegetation-poor still waters with gently sloping banks, e.g., puddles, water-filled lanes, ponds and shallow excavations.

3. Status and decline: In general, this species is moderately frequent in Germany, but rare in the low mountains and extremely rare in the Alps. Declines are moderate to strong in the lowlands, strong in the low mountains, and moderate in the Alps.

4. Red List ranking in Germany: This species is listed as VU in general, but as EN in the low mountains and in the Alps.

5. Protection status: This species is especially and strictly protected; it is listed in Annex IV of the EU's Habitats Directive.

6. Conservation measures: New and restored ponds, availability of terrestrial habitat, regional conservation programmes, and reintroduction projects following captive breeding all provide some protection for this species.

J. *Epidalea calamita*

1. Distribution: The natterjack toad is widespread in Germany and inhabits the dune areas of the northern and Baltic sea coasts, including its islands; however, in the northern lowlands the species is only distributed as scattered populations, and it is lacking altogether at the higher elevations of the low mountains and in the Alps.

2. Habitat: Originally, the natterjack especially inhabited sandy floodplains. Today, primary habitats are found only in coastal dunes and, inland, at places such as heaths. Typical secondary habitats are sand and gravel pits, military training sites and fallow fields with poor vegetation. The waters in which spawning takes place are shallow, temporary, vegetation-free, sun-lit, small ponds or pools, which quickly heat up. Its terrestrial habitats are characterized by dry, warm, loose, sandy and gravelly soils.

3. Status and decline: This species is frequent in Germany, especially in the lowlands and moderately frequent in the low mountains; populations have undergone moderate to strong declines.

4. Red List ranking in Germany: This species is ranked as NT in general, but in the lowlands as LC, and as EN in the mountains.

5. International importance of German populations: Germany contains 10–30% of the distributional range of this species and is centrally located within that range; it therefore comprises an important part of the range this species.

6. Protection status: This species is especially and strictly protected; it is listed in Annex IV of the EU's Habitats Directive.

7. Conservation measures: New and restored ponds, availability of terrestrial habitat, regional conservation programmes, and reintroduction projects following captive breeding all provide some protection for this species.

K. *Hyla arborea*

1. Distribution: The common treefrog is widespread in Germany. It occurs in good numbers, especially in East Germany, where large choruses still are known. It is naturally absent in the marshes of north-western Germany and at the higher elevations of the low mountain ranges (e.g. Harz, Black Forest).

2. Habitat: The original habitats of the treefrog were probably the floodplains of large river systems. A richly structured agricultural landscape in extensively used grasslands, or in wet meadows with high levels of groundwater and composite elements such as hedges, flower-rich margins of fields and forests, represent today's habitat. Small to medium-sized, shallow and sunny still waters with rich herbaceous and shrubby vegetation, such as ponds, pools, oxbows and water in excavations, are used for spawning.

3. Status and decline: This species is moderately frequent in Germany, especially in the lowlands, but very rare in the alpine region. Overall, there are marked declines in populations, but with regional variation; declines are moderate to strong in the lowlands, strong in the low mountains, and strong to very strong in the Alps.

4. Red List ranking in Germany: Overall, this species is ranked as VU, but in the lowlands as NT, and in the mountains and alpine regions as EN.

5. Protection status: This species is especially and strictly protected and is listed in Annex IV of the EU's Habitats Directive.

6. Conservation measures: New installed ponds and restoration of previous ones, availability of terrestrial habitat, regional conservation programmes, and reintroduction following captive breeding contribute to the conservation of this species.

L. *Rana arvalis*

1. Distribution: The western boundary of the moor frog passes through Germany. As a lowland species, its main range is in the north-western and north-eastern lowlands, but it is missing in the alpine region. In central, western and southern Germany the species is scattered in isolated patches, e.g. on the northern plain of the upper Rhine, and in northern Bavaria.

2. Habitat: The moor frog prefers areas with a high water-table (fens, edges of raised bogs, wet grasslands, flood plains, floodplain forests and carrs). Typical sites where spawning occurs are shallow, sunny bodies of standing water that either permanently contain water or temporarily dry out (e.g., ponds, pools, backwaters, oxbows, peat-ditches). For spawning, the moor frog is also able to use acidic waters (such as moor ponds) with pH down to 4.5–5.0.

3. Status and decline: This species is moderately frequent in Germany overall, but frequent in the lowlands and rare in the low mountains. It is strongly in decline in general, with moderate decreases in population size in the lowlands, and very strong ones in the low mountains.

4. Red List ranking in Germany: Overall, this species is listed as VU, but as LC in the lowlands, and as CR in the mountains.

5. International importance of German populations: Owing to its highly isolated populations in the southwest, Germany is a particularly important part of the range of this species.

6. Protection status: This species is especially and strictly protected; it is listed in Annex IV of the EU's Habitats Directive.

7. Conservation measures: New installation and restoration of ponds, restoration of peatlands, and regional conservation programmes contribute to the conservation of this species.

M. *Rana dalmatina*

1. Distribution: The agile frog has a disjunct distribution in Germany and also occurs on some rather large, isolated islands in the south and a few smaller ones in the North German lowlands. As an inhabitant of the planar kollin altitudinal range, the species is naturally rare in the alpine region.

2. Habitat: As a thermophilic species, the agile frog prefers dry, warm, sunny and herb-rich mixed and deciduous forests (European beech; oaks) and their margins, as well as forest meadows. Spawning occurs mainly in smaller, sun-lit, still waters (especially those that are fish-free), mainly, but not always in forests.

3. Status and decline: This species is rare in Germany overall, rare in the low mountains, very rare in the lowlands, and extremely rare in the alpine region; declines of populations are not known.

4. Red List ranking in Germany: This species is listed as LC in general; LC in the lowlands and low mountains; it is rare in the alpine region.

5. International importance of German populations: Owing to the highly isolated populations on the Baltic coast (Mecklenburg-Vorpommern) and in the Lueneburg Heath, Germany is a particularly important part of the range of this species.

6. Protection status: This species is especially and strictly protected; it is listed in Annex IV of the EU's Habitats Directive.

7. Conservation measures: Installation of new ponds and restoration of pre-existing ones and the conversion of forests from coniferous to mixed or deciduous ones, has contributed to the conservation of this species.

N. *Pelophylax lessonae*

1. Distribution: The pool frog does not occur in the northwest and is rare in northern Germany; otherwise it is spread throughout the country. However, its precise distribution is insufficiently known.

2. Habitat: The pool frog prefers fens and wetlands (swamps, alder forests) both within forests and in open countryside. It spawns in permanent, vegetation-rich, sunny, shallow, still waters, such as small ponds in meadows, on floodplains, in woodlands, or at the edges of raised bogs, as well as in temporary ponds and ditches.

3. Status and decline: This species is moderately frequent in Germany generally, mainly in the lowlands, but rare in the alpine region; its extent of decline is unknown but is probably moderate in all regions.

4. Red List ranking in Germany: This species is listed as "Endangering with unknown extent" overall; in the alpine region, it is VU.

5. Protection status: This species is especially and strictly protected; it is listed in Annex IV of the EU's Habitats Directive.

6. Conservation measures: The installation of new ponds and the restoration of pre-existing ones have been beneficial to this species.

III. Conservation measures and monitoring programmes

A. Conservation measures

In Germany, amphibian conservation is the responsibility of the federal states and their nature-conservation authorities. Ministries of the environment are the supreme authorities, with district governments as upper and counties as lower echelons. In addition, there are consultative Nature Conservation Agencies at the national as well as at the federal-state level. These develop, initiate, and implement amphibian conservation programmes and conservation measures. The major part of the protection of amphibians is, however, carried out by volunteers working in national associations for conservation, such as BUND (Bund für Umwelt und Naturschutz Deutschland) and NABU (Naturschutzbund Deutschland) and by many local associations, e.g., Arbeitskreis Amphibien und Reptilien NRW, Amphibien-Reptilien-Biotopschutz Baden Württemberg (ABS), Arbeitsgemeinschaft Amphibienschutz und Reptilienschutz in Hessen (AGAR), Landesverband für Amphibien- und Reptilienschutz in Bayern (LARS), as well as nature conservation foundations, such as Naturstiftung David, Stiftung Naturschutz Schleswig Holstein.

The activities of such groups includes new installations and management of aquatic (restoration; de-sludging; elimination of shading trees and scrub; creation of stepping-stone biotopes) and terrestrial habitats (management of grasslands; connecting habitat by planting hedges and copses), amphibian protection measures on roads, and translocations of endangered populations and reintroductions. The purchase of land for the protection of amphibian habitats is realized by foundations, as well as by national or regional associations for the conservation of nature.

Basic surveys of amphibian populations and derived proposals for their protection, as well as management of habitats are carried out by professional biological/ecological consultants, among others, within the framework of the regulation of impacts (e.g., construction of roads, industrial development, planning land-use) or in the context of the Habitats Directive, reporting duties to the EU, or the designation of nature reserves on behalf of federal, state, district or community authorities. These surveys are, and have been, implemented on a large scale, partly on an international scale. Under the terms of EU-LIFE funding and/or funding by the Federal Ministry for the Environment, projects for the protection and management of *Triturus cristatus, Bombina bombina, Bombina variegata, Alytes obstetricans, Pelobates fuscus, Bufotes viridis, Hyla arborea,* and *Rana arvalis* are carried out by various organizations (mostly nature-conservation associations) in different federal states or even across states. These projects have partly been integrated into projects for the re-naturation of floodplains and raised bogs or industrial and mining landscapes.

At the same time, in various federal states, species' conservation or relief programmes and similar plans for different amphibian species (e.g., see Brandt and Feuerriegel 2004; Sy and Meyer 2004; Geske 2008; Hansbauer and Sachteleben 2008; Ministerium für Ländliche Entwicklung, Umwelt und Verbraucherschutz des Landes Brandenburg, 2009; for examples of websites, see Appendix 1) have been designed and implemented during the past two decades.

B. Monitoring programmes

Systematic monitoring is mainly carried out as part of the Habitats Directive for those species listed in Annexes II and IV (Sachteleben and Behrens 2010, Bundesamt für Naturschutz (BfN) und Bund-Länder-Arbeitskreis (BLAK) FFH-Monitoring und Berichtspflicht 2015).

Mapping distributions and taking inventories takes place particularly at the level of the federal states of Germany, and sometimes also at the municipal level, by voluntary working-groups or advisory authorities for nature conservation, sometimes in cooperation with each other. Extensive handbooks on distribution, ecology, and protection have already been published for many federal states.

For a nationwide Atlas, the working group "field herpetology and species conservation" in the German Society for Herpetology and Herpetoculture (DGHT) was commissioned by the Federal Agency for Nature Conservation in 2014 with the cooperation of the federal state authorities and numerous voluntary associations and working groups, to merge all available data and present them in current distribution maps with corresponding profiles of the species, and to make them accessible to the public (DGHT 2014).

IV. Perspective

Kühnel *et al.* (2009) classified 20% of the amphibian species of the Federal Republic of Germany (*Bufotes variabilis* not yet considered) as very rare to rare, 45% as moderately frequent and 35% as frequent to very frequent. The long-term population trend is rated as very strongly declining to strongly declining for 35% of the species and is considered stable for only 20%. According to Kühnel *et al.* (2009), the Federal Republic of Germany has a special responsibility for the conservation of eight amphibian taxa. The conservation status of the majority of the species listed in the Annexes of the Habitats Directive was generally "unfavourable-bad" or "unfavourable-inadequate"; only three widespread species (*Pelophylax* kl. *esculentus*, *P. ridibundus* and *Rana temporaria*), as well as the less widely distributed *Rana dalmatina*, were ranked as "favourable" and *Pelophylax lessonae* was listed as "unknown" (Ellwanger *et al.* 2014). Unless appropriate agreements are reached between requirements for species' conservation and, above all, agriculture, the number of threatened species and populations is likely to increase further. To further safeguard populations, the potential effects of pathogenic organisms and postulated climatic changes must be systematically investigated.

As the examples from the Red Lists and the Atlas Project have shown, the amphibian monitoring programmes, databases, and conservation programmes and measures, are the responsibility of the federal states. Without questioning this responsibility, a nationwide, central coordination centre for amphibian/reptilian conservation to serve as a nationwide database for monitoring and mapping data in the future, determine the "most important amphibian areas" in Germany, and establish standards for conservation measures, is urgently needed.

V. Acknowledgements

We thank Harold Heatwole and John W. Wilkinson for the certainly time-consuming revision of the English text.

VI. References

Beutler, A., Geiger, A., Kornacker, P.M., Kühnel, K.-D., Laufer, H., Podloucky, R., Boye, P. and Dietrich, E., 1998. Rote Liste der Kriechtiere (Reptilia) und Rote Liste der Lurche (Amphibia) [Bearbeitungsstand: 1997]. In: Binot, M., Bless, R., Boye, P., Gruttke, H. and P. Pretscher. Rote Liste gefährdeter Tiere Deutschlands. *Schriftenreihe für Landschaftspflege und Naturschutz* **55**: 48–52.

Blab, J. and Nowak, E., 1976. Rote Liste der in der Bundesrepublik Deutschland gefährdeten Tierarten. Teil I – Wirbeltiere ausgenommen Vögel (1. Fassung). *Natur und Landschaft* **51**: 34–8.

Böll, S., Tobler, U., Geiger, C.C., Hansbauer, G. and Schmidt, B.R., 2014. Unterschiedliche *Bd*-Prävalenzen und -Befallsstärken verschiedener Amphibienarten und Entwicklungsstadien an einem Chytridpilz belasteten Standort in der bayerischen Rhön. *Zeitschrift für Feldherpetologie* **21**: 183–94.

Brandt, I. and Feuerriegel, K., 2004. Artenhilfsprogramm und Rote Liste Amphibien und Reptilien in Hamburg. Verbreitung Bestand und Schutz der Herpetofauna im Ballungsraum Hamburg. Freie und Hansestadt Hamburg, Behörde für Stadtentwicklung und Umwelt, Naturschutzamt: www.hamburg.de/contentblob/148260/.../rote-liste-amphibien-und-reptilien-textetil.pdf.

Bundesamt für Naturschutz, 1995. Materialien zur Situation der biologischen Vielfalt in Deutschland. Landwirtschaftsverlag, Münster.

Bundesamt für Naturschutz (BfN) und Bund-Länder-Arbeitskreis (BLAK) FFH-Monitoring und Berichtspflicht, 2015. Bewertung des Erhaltungszustandes der Arten nach Anhang II und IV der Fauna-Flora-Habitat-Richtlinie in Deutschland. Überarbeitete Bewertungsbögen der Amphibien und Reptilien als Grundlage für ein bundesweites FFH-Monitoring, Münster: https://www.bfn.de/0315_ffh_richtlinie.html.

DGHT (Ed.), 2014. Verbreitungsatlas der Amphibien und Reptilien Deutschlands, auf Grundlage der Daten der Länderfachbehörden, Facharbeitskreise und NABU Landesfachausschüsse der Bundesländer sowie des Bundesamtes für Naturschutz. www.feldherpetologie.de/atlas.

Ellwanger, G., Symank, A., Buschmann, A., Ersfeld, M., Frederking, W., Lehrke, S., Neukirchen, M., Raths, U., Sukopp, U. and Vischer-Leopold, M., 2014. Der nationale Bericht 2013 zu Lebensraumtypen und Arten der FFH-Richtlinie. Ein Überblick über die Ergebnisse. *Natur und Landschaft* **89**: 185–92.

Franzen, M., Gruber, H.-J. and Heckes, U., 2002. Eine allochthone *Triturus carnifex*-Population in Südbayern (Deutschland). *Salamandra* **38**: 149–54.

Geske, C., 2008. Landesweites Artenhilfskonzept für die Knoblauchkröte in Hessen. *RANA* **5** (Sonderheft): 79–90.

Günther, R. (Ed.), 1996. *Die Amphibien und Reptilien Deutschlands.* Gustav Fischer Verlag, Jena, Stuttgart, Lübeck, Ulm.

Hansbauer, G. and Sachteleben, J., 2008. Das Artenhilfsprogramm Knoblauchkröte (*Pelobates fuscus*) in Bayern. *RANA* **5** (Sonderheft): 91–100.

IUCN, 2017. The IUCN Red List of Threatened Species. Version 2017–3: www.iucnredlist.org. Downloaded on 4 February 2018.

Kühnel, K.-D., Geiger, A., Laufer, H., Podloucky, R. and Schlüpmann, M., 2009. Rote Liste und Gesamtartenliste der Lurche (Amphibia) Deutschlands. Stand Dezember 2008. *Naturschutz und Biologische Vielfalt* **70**: 59–288.

Landesumweltamt Brandenburg (Editor), 2004. Einfluss von Pestiziden auf Laich und Larven von Amphibien am Beispiel eines Herbizides (Isoproturon) und eines Insektizides (Cypermmethrin). *Studien und Tagungsberichte* **49**: 1–104.

Maletzky, A., Mikulíček, P., Franzen, M., Goldschmid, A., Gruber, H.-J., Horák, A. and Kyek, M., 2008. Hybridization and introgression between two species of crested newts (*Triturus cristatus* and *T. carnifex*) along contact zones in Germany and Austria:

morphological and molecular data. *The Herpetological Journal* **18**: 1–15.

Martel, A., Spitzen-van der Sluijs, A., Blooi, M., Bert, W., Ducatelle, R., Fisher, M.C., Woeltjes, A., Bosman, W., Chiers, K., Bossuyt, F. and Pasmans, F., 2013. *Batrachochytrium salamandrivorans* sp. nov. causes lethal chytridiomycosis in amphibians. *Proceedings of the National Academy of Sciences* **110**: 15325–9.

Ministerium für Ländliche Entwicklung, Umwelt und Verbraucherschutz des Landes Brandenburg, 2009. Artenschutzprogramm Rotbauchunke und Laubfrosch. Potsdam: http://www.mlul.brandenburg.de/media_fast/4055/rotbauch.pdf.

Naturschutzverordnung, 1936. Verordnung zum Schutze der wildwachsenden Pflanzen und der nichtjagdbaren wildlebenden Tiere. Reichsgesetzblatt **1**: 181–9, Berlin.

Pätzig, M., 2010. Bedeutung und Biodiversität von Söllen in Nordostdeutschland. *Treffpunkt Biologische Vielfalt* **9**: 83–8.

Ohst, T., Gräser, Y., Mutschmann, F. and Plötner, J., 2011. Neue Erkenntnisse zur Gefährdung europäischer Amphibien durch den Hautpilz *Batrachochytrium dendrobatidis*. *Zeitschrift für Feldherpetologie* **18**: 1–17.

Ohst, T., Gräser, Y. and Plötner, J., 2013. *Batrachochytrium dendrobatidis* in Germany: distribution, prevalences, and prediction of high risk areas. *Diseases of Aquatic Organisms* **107**: 49–59.

Reichsnaturschutzgesetz, 1935. *Reichsgesetzblatt* **1**: 821–25.

Sachteleben, J. and Behrens, M., 2010. Konzept zum Monitoring des Erhaltungszustandes von Lebensraumtypen und Arten der FFH-Richtlinie in Deutschland. BfN-Skripten 278.

Schulte, U. 2014. Artensteckbriefe für alle 20 heimischen Amphibienarten: https://feldherpetologie.de/heimische-amphibien-artensteckbrief/.

SEH 2014: http://www.seh-herpetology.org/Distribution_Atlas/Amphibians_in_Europe_Amphibians.

Sabino-Pinto, J., Bletz, M., Hendrix, R., Perl, R.G.B., Martel, A., Pasmans, F., Lötters, S., Mutschmann, F., Schmeller, D.S., Schmidt, B.R., Veith, M., Wagner, N., Vences, M. and Steinfartz, S. (2015): First detection of the emerging fungal pathogen *Batrachochytrium salamandrivorans* in Germany. *Amphibia-Reptilia* **36**: 411–6.

Sillero, N., Campos, J., Bonardi, A., Corti, C., Creemers, R., Crochet, P.-A., Crnobrnja-Isailovic, J., Denoël, M., Ficetola, G.F., Gonçalves, J., Kuzmin, S., Lymberakis, P., de Pous, P., Rodríguez, A., Sindaco, R., Speybroeck, J., Toxopeus, B., Vieites, D.R. and Vences, M., 2014. Updated distribution and biogeography of amphibians and reptiles of Europe. *Amphibia-Reptilia* **35**: 1–31.

Spitzen-van der Sluijs, A., Martel, A., Asselberghs, J., Bales, E.K., Beukema, W., Bletz, M.C., Dalbeck, L., Goverse, E., Kerres, A., Kinet, T., Kirst, K., Laudelout, A., Marin da Fonte, F., Nöllert, A., Ohlhoff, D., Sabino-Pinto, J., Schmidt, B.R., Speybroeck, J., Spikmans, F., Steinfartz, S., Veith, M., Wagner, N., Pasmans, F. and Lötters, S., 2016. Expanding distribution of lethal amphibian fungus *Batrachochytrium salamandrivorans* in Europe. *Emerging Infectious Diseases* **22**: 1286–8.; https://dx.doi.org/10.3201/eid2207.160109.

Statistisches Bundesamt 2017: Statistisches Jahrbuch 2017: https://www.destatis.de/DE/Publikationen/StatistischesJahrbuch/StatistischesJahrbuch2017.html.

Stöck, M., Roth, P., Podloucky, R. and Grossenbacher, K. (2008): Wechselkröten unter Berücksichtigung von *Bufo viridis viridis* Laurenti, 1768; *Bufo variabilis* (Pallas, 1769); *Bufo boulengeri* Lataste, 1879; *Bufo balearicus* Böttger, 1880 und *Bufo siculus* Stöck, Sicilia, Belfiore, Lo Brutto, Lo Valvo & Arculeo, 2008. In: Grossenbacher, K. (Hrsg.): *Handbuch der Reptilien und Amphibien Europas – Froschlurche II. – Volume 5/II*: 413–98.

Sy, T. and Meyer, F., 2004. Bestandssituation und Schutz der Rotbauchunke in Sachsen-Anhalt. Fachteil zum Artenhilfsprogramm. Berichte des Landesamtes für Umweltschutz Sachsen-Anhalt, Sonderheft 3.

Appendix 1. Species' conservation or relief programmes (selection)

Bayerisches Landesamt für Umwelt – Artenhilfsprogramm Amphibien. http://www.lfu.bayern.de/natur/artenhilfsprogramme_zoologie/amphibien/index.htm.

Vollzugshinweise für Arten und Lebensraumtypen. http://www.nlwkn.niedersachsen.de/naturschutz/natura_2000/vollzugshinweise_arten_und_lebensraumtypen/vollzugshinweise-fuer-arten-und-lebensraumtypen-46103.html.

56 Conservation and declines of amphibians in Poland

Maciej Pabijan and Maria Ogielska

Abbreviations or acronyms used in the text or references

EPMAC	*Educative and Participative Monitoring for wider and more effective Amphibian Conservation*
GIOŚ	*Główny Inspektorat Ochrony Środowiska (Chief Inspectorate of Environmental Protection)*
IUCN	*International Union for the Conservation of Nature*
Kya	*thousand years ago*
m a.s.l.	*metres above sea level*
MP	*Maciej Pabijan*

I. Introduction

A. Climate and landscape

Poland is a central European country of medium size (312,679 km²) located between the Baltic Sea in the north and the Carpathian and Sudetes Mountains in the south. The topography between these major geographic features consists mainly of plains and low hills in some eastern and south-central regions, usually not exceeding 400 m a.s.l. Only the southern part of Poland, i.e. 8.7% of the total surface area, lies above 300 m a.s.l. (6% of which includes the north-eastern fringe of the Carpathian mountain range) with several dozen peaks exceeding 2000 m a.s.l. in the Tatra Mountains on the border with Slovakia.

The climate in Poland (Woś 1999) is warm summer continental (Dfb, according to the Köppen classification of climate), milder in the northwestern provinces owing to the influence of the Baltic Sea, but becoming more continental towards the east and especially southeast, with hotter summers and colder winters. Mean annual temperature is rather uniform and ranges from 7° to 9°C (excluding higher elevations). Rainfall is variable with lowest values in the central provinces (500–600 mm), and higher precipitation near the Baltic Sea (up to 800–900 mm). Annual precipitation across the northern (Polish) Carpathians and Sudetes reaches 1000–1500 mm.

The surface of the northern third of the country is geologically young (Würm glaciation) and contains a large number of lakes, marshes, ponds, peat bogs, river floodplains and oxbow lakes. The landscape of most of central and south-central Poland is geologically older, more homogeneous, and has been transformed by large-scale intensive agriculture and industry during the 20th century.

Differences in orography and climate determine two biomes in Poland: the Continental Biome extending over the majority of the country, and the Alpine Biome limited to the Carpathians and Sudetes together with their forelands. The natural vegetation of the Continental Biome includes deciduous oak, hornbeam, lime, beech and alder forests, with stands of spruce in the northeast. The Alpine Biome (Figure 56.1) was originally covered by beech and silver fir, with sycamore in the ravines. Silviculture has drastically changed the composition of forests by replacing natural

Fig. 56.1 Map of Poland showing areas above 300 m a.s.l. (shaded grey), remaining forest cover (green), national parks (hatched orange) and landscape parks (hatched yellow). A dotted red line delimits the northern boundary of the Alpine biome. Numbers mark the approximate locations of sites to which reference is made in the text. They are from northwest to southeast: 1 – Słupsk, 2 – Gdańsk, 3 – Olsztyn, 4 – Wigierski National Park, 5 – Biebrzański National Park, 6 – Narwiański National Park, 7 – Białystok, 8 – Warszawa, 9 – Gniezno, 10 – Poznań, 11 – Zielona Góra, 12 – Barycz Valley, 13 – Wrocław, 14 – Oława, 15 – Wałbrzych, 16 – Piotrków Trybunalski, 17 – Katowice, 18 – valleys of upper Vistula, lower Sola and Skawa rivers, 19 – Ojców National Park, 20 – Kraków, 21 – Niepołomice Forest, 22 – Nida Basin (Bonk and Pabijan 2010), 23 – Nowy Sącz, 24 – Poleski National Park, 25 – Beskid Niski, 26 – Bieszczady.

formations with stands of Scots pine in the lowlands and spruce in mountainous areas (Łonkiewicz 1996). The State Forests National Forest Holding (hereafter: State Forests), an organization utilizing and protecting publicly owned forests in Poland, estimates that the percentage of land currently covered by forest in Poland is 29.5% (www.lasy.gov.pl) (Figure 56.1).

B. Overview of amphibian diversity

Poland is inhabited by at least 19 species of amphibians (Table 56.1), constituting 24% of the total number of native species reported for Europe (Speybroeck *et al.* 2010). The species' checklist includes five caudates, 13 anurans and one natural hybridogenetic species, *Pelophylax* kl. *esculentus*, discovered by a Polish zoologist, Prof. Leszek Berger in the 1960s. There are no species endemic to Poland; however, *Lissotriton montandoni* is restricted to the Carpathians and is considered a central European endemic. The Carpathian populations of *Bombina variegata* are also genetically distinct from southern and western populations (Szymura *et al.* 2000; Spolsky *et al.* 2006).

The Continental and Alpine biomes are inhabited by distinct amphibian assemblages. The Alpine assemblage consists of six species (*Rana temporaria, Bufo bufo, Bombina variegata, Salamandra salamandra, Ichthyosaura alpestris* and *Lissotriton montandoni*) and occurs in the Carpathian and Sudetes mountain ranges of southern Poland. The Continental assemblage inhabits the lowlands north of the Carpathians and Sudetes and is composed of 14 taxa (*Rana temporaria, R. arvalis, Bufo bufo, Bufotes viridis* (including *B. variabilis*), *Epidalea calamita, Bombina bombina, Hyla arborea/H. orientalis, Pelobates fuscus, Pelophylax lessonae, P. ridibundus, P.* kl. *esculentus, Triturus cristatus* and *Lissotriton vulgaris*). There are two cases of species-pairs in which each member of a pair occupies

Table 56.1   Status of amphibians in Poland. IUCN Red List/Polish Red Data Book status (Głowaciński 2001; RDB). LC – least concern, NT – near threatened, NE – not evaluated. Legal status: legal protection of species in Poland (J. of Laws of 2016, item 2183): + / + fully protected, + / – partial protection. Monitoring GIOŚ: results of two rounds of surveys at 5–8 year intervals for species for which nationwide monitoring programs have been implemented by the Chief Inspectorate of Environmental Protection; – refers to species not included in monitoring or for which time-series data are not yet available. Trend SC Poland: gives the trend in number of amphibian populations in south-central Poland (Nida Basin) over a 25-year period, according to Bonk and Pabijan (2010). Asterisks reflect significant trends after stringent statistical criteria. na – not present in study area or not assessed due to low sample sizes.

Family	Genus	Species	Common name	IUCN/RDB	Legal status	Monitoring GIOŚ	Trend SC Poland
Salamandridae	*Triturus*	*cristatus*	great crested newt	LC/NT	+ / +	decline	decline*
Salamandridae	*Lissotriton*	*vulgaris*	smooth newt	LC/NE	+ / –	–	decline*
Salamandridae	*Lissotriton*	*montandoni*	Carpathian newt	LC/LC	+ / +	–	na
Salamandridae	*Ichthyosaura*	*alpestris*	alpine newt	NE/NE	+ / –	–	na
Salamandridae	*Salamandra*	*salamandra*	fire salamander	LC/NE	+ / –	–	na
Bombinatoridae	*Bombina*	*bombina*	fire-bellied toad	LC/NE	+ / +	decline	decline
Bombinatoridae	*Bombina*	*variegata*	yellow-bellied toad	LC/NE	+ / +	–	na
Pelobatidae	*Pelobates*	*fuscus*	common spadefoot	LC/NE	+ / +	decline	increase
Bufonidae	*Bufo*	*bufo*	common toad	LC/NE	+ / –	–	decline
Bufonidae	*Bufotes*	*viridis*	green toad	LC/NE	+ / +	stable	stable
Bufonidae	*Epidalea*	*calamita*	natterjack toad	LC/NE	+ / +	decline	na
Hylidae	*Hyla*	*arborea/orientalis*	European tree frog	LC/NE	+ / +	decline	stable
Ranidae	*Rana*	*temporaria*	common frog	LC/NE	+ / –	decline	decline*
Ranidae	*Rana*	*arvalis*	moor frog	LC/NE	+ / +	decline	stable
Ranidae	*Rana*	*dalmatina*	agile frog	LC/NT	+ / +	–	na
Ranidae	*Pelophylax*	*lessonae*	pool frog	LC/NE	+ / –	decline	decline*
Ranidae	*Pelophylax*	kl. *esculentus*	edible frog	LC/NE	+ / –	stable	decline
Ranidae	*Pelophylax*	*ridibundus*	marsh frog	LC/NE	+ / –	increase	increase*

either the Alpine or Continental climatic zone: *Lissotriton vulgaris/L. montandoni* and *Bombina bombina/B. variegata*. Two species, *R. temporaria* and *B. bufo*, are widely distributed in both bioclimatic regions.

The local ranges of the Alpine and Continental amphibian species overlap in the foothills and river valleys north of the Carpathians and Sudetes. The diverse landscape of these regions includes habitat patches suitable for both lowland and highland amphibians, and potentially 18 species may be found within a few kilometres of each other. This area also holds populations of *Rana dalmatina*, the rarest amphibian in Poland with a fragmented and understudied distribution (Bonk *et al.* 2012). Hybrid zones between *Bombina bombina* and *B. variegata* (Szymura 1993), as well as between *Lissotriton vulgaris* and *L. montandoni* (Babik *et al.* 2003), are known from the foothills and lower Carpathian ranges. Unfortunately, a dense road network, urbanization (including sprawling agglomerations such as Kraków and Katowice) and agriculture have transformed the landscape, thereby reducing and fragmenting amphibian habitats.

C. Phylogeographic patterns

At the height of the Pleistocene glaciations, Poland was almost certainly uninhabited by ectothermic vertebrates, including amphibians, owing to a harsh climate and a succession of ice sheets that at times covered most of the territory. The lack of cold-stage refugia for ectotherms implies that all current amphibian species re-colonized the territory after the retreat of glaciers and tundra (after *ca.* 11.7 kya). This is reflected in typically low levels of genetic diversity in those species that have been studied in detail, e.g. *Bombina bombina* (Hofman *et al.* 2007; Fijarczyk *et al.* 2011), *Ichthyosaura alpestris* (Pabijan *et al.* 2005; Pabijan and Babik 2006) and *Triturus cristatus* (Babik *et al.* 2009).

The phylogeographic patterns of European amphibians indicate that the territory of Poland, and of north-central Europe in general, is a crossroads for lineages that have expanded from several Pleistocene refugia. These refugia include the Balkans (*Lissotriton vulgaris*) (Babik *et al.* 2005; Pabijan *et al.* 2015, 2017); *Salamandra salamandra* (Najbar *et al.* 2015; Konowalik *et al.* 2016), southern Carpathians (*Bombina variegata, Lissotriton montandoni*; Hofman *et al.* 2007; Fijarczyk *et al.* 2011; Zieliński *et al.* 2013), Pannonian Basin (*Triturus cristatus*; Wielstra *et al.* 2013), lowlands near the Black Sea (*Bombina bombina* and *Pelobates fuscus* (Hofman *et al.* 2007; Fijarczyk *et al.* 2011; Litvinchuk *et al.* 2013), southwestern Europe (*Ichthyosaura alpestris*; Pabijan and Babik 2006; Sotiropoulos *et al.* 2007; Recuero *et al.* 2014), and southern Russia (*Rana arvalis*) (Babik *et al.* 2004). Two cryptic species of treefrog, *Hyla arborea* and *H. orientalis* (Stöck *et al.* 2012) have colonized Poland from the south (potential refugia in the Balkan Peninsula) and southeast (Black Sea), respectively. Contact and hybrid zones between the two species of treefrog are located meridionally, roughly in the centre of the country along the Vistula River (Stöck *et al.* 2012; Dufresnes *et al.* 2016). Two mitochondrial lineages of the green toad provisionally identified as *Bufotes viridis* and *B. variabilis* by Stöck *et al.* (2006) are distributed over much of central Poland and occur syntopically at some sites (Maciej Pabijan, unpublished data).

D. Aquatic and terrestrial habitats

Amphibians are closely associated with aquatic habitats, which are usually regarded as the main sites in need of conservation. However, the surrounding terrestrial habitat is equally important as it provides foraging and wintering grounds for many species. Moreover, dispersal among habitat patches and breeding sites depends on the quality of intervening terrestrial habitat. The risk of extinction is buffered through the metapopulation structure of many amphibian species by maintaining a network of ponds and the connecting habitats between them, including buffer zones (Semlitsch 2003). Although it is clear that bodies of water cannot be managed separately from surrounding terrestrial habitat, the latter is often overlooked in conservation-oriented research, landscape planning or practical programmes aimed at conserving amphibians in Poland (Makomaska-Juchiewicz and Baran 2012).

All but one amphibian species (*Salamandra salamandra*) in Poland depend on lentic water for reproduction. Common amphibian breeding sites include small ponds in forests or agricultural landscapes, temporary or permanent bodies of water located along riverine floodplains, as well as marshes, alder swamps and other wetlands. Arguably, the most intact amphibian habitat in the country is located in the northeastern lowlands and the Lake District, a sparsely populated, densely wooded area subject to traditional extensive farming. In some regions, the density of amphibian breeding sites is high, as in the Biebrzański, Wigierski and Narwiański National Parks (Adaros *et al.* 2000) and in the Poleski National Park in the southeast (Figure 56.1) (Czarniawski *et al.* 2014). Large and viable amphibian metapopulations probably exist in these areas.

In the Alpine Biome, natural breeding sites include water-filled basins originating from landslides and the erosion of slopes, high-elevation meltwater ponds and water-filled depressions along stream valleys. Important breeding sites in highland areas include temporary bodies of water (e.g. large puddles, wheel-ruts, floodplains) that are typically present from March until July or even until mid-August, covering the period from spawning to the completion of metamorphosis. In the southeast, the heavily forested hills of the Beskid Niski and Bieszczady ranges are characterized by gentle slope, clayey, water-retaining soil and ample rainfall, thereby providing numerous breeding sites and terrestrial habitat for all alpine amphibian species, and at lower elevations, other species such as *T. cristatus*, *L. vulgaris*, *R. dalmatina* and *H. arborea* (Świerad 2003; Głowaciński and Sura 2018). The larviparous *Salamandra salamandra*, found exclusively in the Alpine Biome, is the only amphibian in Poland that regularly uses small streams for deposition of larvae, although small ponds and other lentic waters are also occasionally used by this species (Zakrzewski 2007).

Man-made bodies of water, such as abandoned sand, gravel or stone quarries and peat-excavation pits are commonly used as breeding sites by amphibians. Forestry practice contributes to the creation of temporary water basins such as wheel-ruts and water-filled depressions produced by timber extraction. In some areas these man-made structures constitute the predominant amphibian breeding sites (Babik and Rafiński 2001; Pabijan *et al.* 2009; Sadza *et al.* 2015). In the highly industrialized regions of south-central Poland, large expanses of overburden or other waste rock removed during mining has created an irregular landscape with water-filled potholes and depressions that support amphibian breeding aggregations (Tomalka-Sadownik and Rozenblut-Kościsty 2010). Flooded stone or gravel quarries may also harbour diverse and numerous amphibian communities (Wirga and Majtyka 2015; Klimaszewski *et al.* 2016). In other areas where pristine wetlands have been destroyed, semi-natural carp ponds stocked with different age-classes of fish function as substitute breeding sites for several amphibian species (Kloskowski 2010). Some of the most biodiverse and extensive complexes of fish ponds were created in the Middle Ages and are protected as Birds or Habitats Directive Sites in the Natura 2000 framework. These include ponds and floodplains in the valley of the Barycz River (PLB020001, PLH020041) and an extensive carp husbandry centre in the valleys of the upper Vistula (PLB240001, PLB120009), lower Sola (PLB120004) and lower Skawa rivers (PLB120005) of south-central Poland. These sites contain a diverse and abundant amphibian fauna. For this reason Berger (2008) proposed to maintain or create amphibian breeding ponds, along with regular carp ponds in fishery farms. A similar proposal was suggested by Ogielska and Kierzkowski (2010) as "amphibian asylums", i.e. small bodies of water close to recreation ponds.

Amphibians and their habitats in the urban environment have gained attention in recent years and were the focus of a special issue of Fragmenta Faunistica (vol. 53, No. 2, 2010) including nine cities (Zielona Góra, Wałbrzych, Wrocław, Oława, Gniezno, Słupsk, Olsztyn, Białystok and Lublin).

Most species inhabiting Poland (excluding *Salamandra salamandra, Lissotriton montandoni, Bombina variegata* and *Rana dalmatina*) were found in the cities, and the average number of species was 10.6 (range: 8–13) (Mazgajska and Mazgajski 2010).

II. Threats to amphibians in Poland

A. Destruction of aquatic and terrestrial habitats

The destruction or degradation of aquatic breeding habitats has been singled out as the main contributor to amphibian population declines in Poland in the 20th century (Berger 1987; Młynarski 1987; Głowaciński and Rafiński 2003). The landscape of Poland was severely impacted by a centrally planned intensive agricultural policy of the communist regime during the second half of the 20th century. Large-scale drainage of natural wetlands attained highest levels in the latter half of the century (Ciepielowski and Gutry-Korycka 1993) with 20% of the land being affected. Intervention in the natural courses of rivers led to the regulation of 50,000 km of rivers and canals (Mioduszewski 1997). These activities took a large toll on amphibian breeding sites in Poland through a drop in groundwater level and through deliberate draining and filling of a large proportion of small ponds, marshes and wetlands. Stasiak (1991; after Rybacki and Berger 2003) reported a drop of 77.5% in the number of small ponds in Wielkopolska (western Poland) between the 1890s and 1960s. Pieńkowski (2003) reported a 70% decrease in the number of small ponds in West Pomerania province (north-western Poland) between the end of the 19th and the second half of the 20th century. Rybacki and Berger (2003) reported a loss of 19% of small ponds in the agricultural landscape of western Poland (Wielkopolska) over a 30–40 year period in the second half of the 20th century. An even greater decrease of 59% was noted for central Poland (near Piotrków Trybunalski) by Olaczek *et al.* (1990) over a similar period of time.

More recently, increasing urbanization has also reduced amphibians' habitats. For instance, in the vicinities of Zielona Góra (south-western Poland), Najbar *et al.* (2005) noted that 40% of amphibian-inhabited reservoirs were drained or filled in between 1974 and 2004, mostly owing to encroaching housing development. Najbar *et al.* (2017) also noted loss of habitat and increased mortality in an urban population of *Salamandra salamandra*. Fortunately, not all data are so alarming. An inventory of various kinds of natural and anthropogenic bodies of water in Wrocław (Nawara *et al.* 2005) revealed a similar number over a 40-year period (211 bodies of water in 1966 versus 195 in 2005).

Another threat to amphibian habitats in Poland is the degradation of water quality due to runoff of mineral and organic fertilizers (Berger 1987, 1989; Rybacki and Berger 2003). Agrochemicals may induce mortality either directly or may increase the rate of eutrophication and therefore also increase the rate of overgrowing. Unfortunately, small ponds are also commonly used as refuse dumps or for disposal of sewage (Rybacki and Berger 2003). Many breeding sites lying close to villages and cities are polluted by chemicals and garbage, but only a few are documented (Budzik *et al.* 2014; Kolenda *et al.* 2014). Pollution may also induce bone anomalies in some species: Kaczmarski *et al.* (2016) showed that 80% of adult toads (*Bufo bufo*) from urban areas have phalangeal malformations, compared to 20% of individuals from rural or semi-urban areas.

We also note that, despite a number of national and EU legal provisions in place for the protection of amphibian species and their environment, and the dedicated efforts of conservationists (in particular the outstanding work of Marek Sołtysiak), entire populations and habitats are still being decimated owing to infrastructure, industry or housing developments. Ecological impact assessments (required by law) are often inadequate or erroneous (Sołtysiak 2006; Sołtysiak and Matusiak 2006, Sołtysiak and Kaźmierczak 2008; Sołtysiak and Rybacki 2010) and are rarely reviewed by independent experts. The effectiveness of compensatory mitigation to replace lost aquatic and terrestrial habitat is likewise seldom satisfactory (Sołtysiak and Rybacki 2012). The majority of landscape-management decisions, including actions that destroy or adversely modify critical habitat for protected wildlife, are made at Regional Directorates for Environmental Protection. Unfortunately, administrators at these institutions often lack either knowledge or willpower to effectively protect the environment or, if informed of potential threats to amphibian populations, demonstrate indolence instead of a genuine attempt at mitigation (Sołtysiak and Rybacki 2010).

B. Mortality on roads

With continuing investment in the construction of highways and an increasing volume of traffic, amphibian mortality on roads has become a pressing conservation issue affecting amphibians in Poland. Collisions with automobiles are particularly acute during the spring breeding season (Orłowski *et al.* 2008) but may also be high during autumnal migrations (Gryz and Krauze 2008; Wojdan 2010). Mortality of adults on roads was identified as a key parameter governing the metapopulation dynamics of natterjack toads and their long-term conservation in central Poland (Franz *et al.* 2013). Amphibians' mortality on roads has been well-documented in suburban areas (Najbar *et al.* 2006; Hetmański *et al.* 2007; Elzanowski *et al.* 2009; Błażuk 2010), agricultural landscapes (Orłowski 2007), and in relatively natural areas such as national parks or other forms of protected landscape (Figure 56.1) (Wołk 1978; Rybacki 1995; Zamachowski and Plewa 1996; Bartoszewicz 1997; Baldy 2002; Rybacki and Krupa 2002; Rybacki and Domańska 2004; Gryz and Krauze 2008; Ogielska *et al.* 2008; Elzanowski *et al.* 2009; Wojdan 2010; Brzeziński *et al.* 2012; Arciszewski 2015). Apart from documentation of the threat, there have been critical assessments of the methods used to estimate the impact of road traffic on nearby amphibian populations (Elzanowski *et al.* 2009; Brzeziński *et al.* 2012) and attempts to identify variables of the habitat that affect mortality (Orłowski 2007; Orłowski *et al.* 2008). Amphibian mortality caused by rail traffic (trains, trams) is low and seems to be less of a threat than collisions with automobiles (Budzik and Budzik 2014; Kaczmarski and Kaczmarek 2016).

C. Introduction of predatory and exotic species

The stocking of adult fish or fry into medium-sized or small ponds for subsistence or recreational purposes is a common practice in Poland. The release of predatory species that feed on tadpoles and caudate larvae into ponds that previously did not harbour populations of fish may be a serious but underestimated threat to amphibians in some areas (Rybacki and Maciantowicz 2006; Bonk and Pabijan 2010). For instance, in Niepołomice Forest, 10 of 20 small ponds surveyed for amphibians contained individuals of *Esox lucius*, *Carassius* sp., or *Perca fluviatilis* (M. Pabijan, unpublished). These ponds are not connected to any drainage system and fish have most likely been intentionally introduced along with excessive vegetation removed during the maintenance of nearby drainage ditches and canals (information from Forestry officials). Unfortunately, explosively breeding amphibians, especially *Rana temporaria* and *Bufo bufo*, are commonly believed to be a threat to fish, and the owners of fish ponds were even encouraged by authors of textbooks devoted to the maintenance of fish ponds (e.g. Guziur 2004) to kill, destroy and remove egg masses and tadpoles.

An invasive species of fish, the Amur sleeper (*Perccottus glenii*) has recently (first record in 1993) colonized the drainage system of the Vistula and Oder rivers (Witkowski 2011). This ecologically versatile species is a voracious predator of invertebrates and small vertebrates (including amphibian larvae) and has decimated amphibian populations in western Russia (Reshetnikov 2003). The alarming rate of expansion of this species in the Vistula drainage basin (80–100 km/per year)

(Witkowski 2011) leads us to flag this invasion as a potential threat to amphibian species in Poland over the next decade.

Invasive amphibian species, such as the American bullfrog (*Lithobates catesbeianus*) or African clawed frog (*Xenopus laevis*), have not been reported. Recently, Kolenda *et al.* (2017b) provided genetic evidence for an admixture of native *Pelophylax* frogs and the non-native *Pelophylax kurtmuelleri*, a predominantly Balkan taxon, in south-western Poland (Barycz River drainage system). These frogs contained mitotypes identical with those of frogs from northern Greece and Albania, suggesting hybridization following human-mediated introduction of *P. kurtmuelleri*, although alternative explanations, e.g. retention of ancestral polymorphism or recent expansion of admixed *P. kurtmuelleri*, could not be ruled out.

D. Infectious diseases

The distribution and prevalence of infectious diseases such as chytridiomycosis (caused by *Batrachochytrium* sp.) or ranavirus (Iridoviridae) in Poland, and their potential impact on amphibian populations, are not well known. The chytrid fungus *B. dendrobatidis* has been detected in *Bombina variegata*, *Pelophylax lessonae* and *P. kl. esculentus* collected from several localities (Sura *et al.* 2010; Kolenda *et al.* 2017a). Individuals of *P. lessonae* from fish ponds in southwestern Poland had extremely high levels of infection that may be linked to reduced fitness and mortality in this population (Kolenda *et al.* 2017a). We note that *B. dendrobatidis* is widespread in neighbouring Germany and in the Czech Republic (Civiš *et al.* 2012; Ohst *et al.* 2013) but no population declines attributed to chytridiomycosis have been reported. Waterfrogs (*Pelophylax* spp.) co-infected with *B. dendrobatidis*, ranavirus, and bacteria causing red-leg disease were found in two fishponds near the Oder river in southwestern Poland (K. Kolenda, pers. communication). Recently, a survey of 13 populations of *Salamandra salamandra* did not reveal any cases of infection by *B. salamandrivorans* (K. Kolenda, personal communication). Clearly, more field surveillances are urgently required, especially in the wake of the expansion of *B. salamandrivorans* (Spitzen-van der Sluijs *et al.* 2016). This state of affairs has been noted by nature conservation institutions that have issued recommendations aimed at restricting the spread of chytridiomycosis and other amphibian diseases, and that have appealed for increased monitoring of susceptible amphibian populations and regulations in trade in exotic species (Państwowa Rada Ochrony Przyrody/2/2018; Generalny Dyrektor Ochrony Przyrody DZP-WG.605.05.3.2018.ep.2).

E. Climatic change

Temporal trends in spawning times of two early-breeding anurans were recorded over a 25-year period in western Poland (Wielkopolska), showing that first spawning has advanced by nine days for *Bufo bufo*, with a similar but non-significant trend for *Rana temporaria* (Tryjanowski *et al.* 2003). These trends appear to be linked to increasing winter and spring temperatures (especially in March) in the study area. It is unknown whether the phenological changes have affected the survival rates or population dynamics of these species. One study suggested a possible link between a progressively milder climate in western Poland and changes in morphology of frogs: an increase in body size in *Pelophylax lessonae* and *P. ridibundus* over a 40-year period was positively correlated with the winter North Atlantic Oscillation; however, the opposite trend was found for *P. kl. esculentus* (Tryjanowski *et al.* 2006).

III. Evidence for amphibian declines

Anecdotal evidence indicates that amphibians in Poland have declined in recent decades (Głowaciński *et al.* 1980; Berger 2000; Głowaciński and Rafiński 2003). Most contemporary Polish herpetologists would agree that loss and degradation of aquatic habitat has been the main factor behind the declines (Rybacki and Berger 2003), and many have observed a drop in the number of individuals in populations that have not disappeared. However, there is a dearth of quantitative evidence for amphibian declines. There are only a few assessments of long-term trends for amphibian populations, i.e. studies that recorded absolute numbers of individuals. Rybacki and Berger (2003) recorded a decline in common toads, common frogs and moor frogs in a single pond in western Poland (Wielkopolska) between 1977 and 2002, while the numbers of water frogs fluctuated but did not show a net loss over this period. Berger (1987, 1989) attributed the declines at this and other sites in the vicinity to excessive levels of nitrogenous fertilizers resulting in the death of larvae. In southern Poland (Niepołomice Forest), the abundance of anurans in a terrestrial, forested ecosystem has declined at least three-fold over a 50-year period. In 1967–1968, Głowaciński and Witkowski (1970) used removal sampling and estimated an average of 2044 individuals/ha (range 1555–2288, from five experimental plots) inhabiting the floor of this deciduous forest. The same area held on average 513 individuals/ha (range 33–1344, from 17 experimental plots) in 2016 and 2017 (M. Pabijan and M. Bonk, unpublished). Anuran biomass showed a similar pattern, declining from 11560 g/ha (range 8278–13290 g/ha) in 1967/68, to 3909 g/ha (range 217–8795 g/ha) half a century later. These results suggest a substantial decline of amphibians in a relatively natural forest complex. Potential causes include industrial pollution, mortality associated with increased traffic on a nearby road, and a general drop in ground-water levels limiting the number and extent of breeding sites. At another site in southern Poland, Schimscheiner (1990) detected a three-fold decline in numbers of *Bombina bombina* between 1973 and 1988.

Surveys of multiple breeding sites have invariably compared only two points in time (usually over 10–25 year intervals). In terms of spatial scale, the most comprehensive assessment of amphibian population status recorded presence/absence data for 71 breeding sites in south-central Poland (Nida Basin) (Bonk and Pabijan 2010) that were previously surveyed 25 years earlier (Juszczyk *et al.* 1988, 1989). On average, the number of species per locality declined by 2.2, and the number of populations was lower for seven species (Table 56.1). Declines in four species (*T. cristatus*, *L. vulgaris*, *R. temporaria* and *P. lessonae*) were severe, showing losses of over 30% of populations in this area (Bonk and Pabijan 2010). Interestingly, the number of localities for *Pelophylax ridibundus* increased from 2 to 17, a finding attributed to this species' mobility, tolerance for breeding in waters inhabited by predatory fish, as well as its use of rivers and irrigation canals as avenues for dispersal (although misidentifications of *Pelophylax* taxa could have confounded these results). The elimination of wetlands due to drainage and the filling of depressions in the agricultural landscape may have eliminated breeding sites and increased inter-pond distances in this area. Deliberate introduction of predatory fishes, an increase in the number of paved roads, and heavier volume of traffic may have all contributed to declines in this area.

In Poland, the distribution of *Salamandra salamandra* reaches its northeastern limit along the border of the Sudetes and Carpathians mountains. During the past 30 years, about 30% of populations of this species have disappeared, mainly in the western part of the Sudetes, and as a result the northern edge of its distribution has shifted southwards by about 20 km (Ogrodowczyk *et al.* 2010). The most probable cause for this decline is destruction of habitat due to urbanization, construction of roads and deforestation. The fire salamander has also declined at its upper elevational limit in the Tatra National Park of southern Poland (Sadza *et al.* 2015).

Several smaller-scale studies are also available. Szyndlar (1994/1995) noted lower numbers of breeding populations of six species (*L. vulgaris*, *T. cristatus*, *Bufo bufo*, *Bombina bombina*, *Bufo viridis* and *Hyla arborea*) in Ojców National Park over a period of 11 years (1977–1988), and attributed the declines to loss and pollution of aquatic habitat. Only the number of breeding populations of *R. temporaria* remained stable over this period. In the vicinities of Zielona Góra (southwestern Poland),

Najbar *et al.* (2005) noted the local extirpation of *Epidalea calamita*, *Hyla arborea* and *Bombina bombina* due to habitat destruction between 1974 and 2004. A loss of amphibian habitat and breeding populations of six amphibian species were also noted for urban and suburban areas of Kraków (Budzik *et al.* 2013). In contrast to previous studies, Kaczmarek *et al.* (2014) described a slight (and nonsignificant) decrease in mean number of amphibian species per pond in the city of Poznań over a 20-year period. They speculated that their lack of evidence for a decline, despite continuing fragmentation of habitat and increased intensity of road traffic, is due to a time-lag between changes in the urban environment and species' occurrence (Löfvenhaft *et al.* 2004).

IV. Inventory and monitoring

Important historical contributions on the distributions of amphibians in Poland include Berger *et al.* (1969) and Juszczyk (1987). The systematic recording of amphibian distributions based both on published inventories and on unpublished records began in 1988 at the Institute of Nature Conservation (Polish Academy of Sciences) in Kraków. This culminated in the Atlas of the Amphibians and Reptiles of Poland (Głowaciński and Rafiński 2003) in which 30,000 records, covering about 2/3 of the area of the country, were evaluated and mapped. Since this time there has been a marked increase in the number of herpetological surveys and inventories, many of them carried out by a new generation of more numerous researchers and NGOs. An initiative to revise the existing atlas and produce a second edition has been completed and is in press (Głowaciński and Sura 2018). This revised atlas is based on 65,500 records covering 90% of the country and includes a regularly updated online edition (http://www.iop.krakow.pl/plazygady).

Recently, an atlas project has documented the distribution of amphibian breeding sites within the city of Poznań (Kaczmarek *et al.* 2014). Similar plans exist for Gdańsk. These small-scale but detailed inventories, headed by NGOs, are aimed at directly supplying (as GIS layers) high-quality, up-to-date information on the local distributions of amphibians for use by regional governmental institutions in urban planning.

Poland has a legal obligation to monitor the conservation status of natural habitats and species with particular regard to priority natural habitat types and priority species in Annex I and Annex II of the EU Habitats Directive. The obligation to carry out such monitoring was imposed by national law (beginning with J. of Laws of 2004, item 880 with later amendments; recent updates in J. of Laws of 2016, item 2183), legislation by the European Union, and a number of international conventions, in particular the Convention on Biological Diversity. The Chief Inspectorate of Environmental Protection has commissioned the monitoring of habitats and species by the Institute of Nature Conservation of the Polish Academy of Sciences in Kraków. For amphibians, the first national monitoring program was initiated in 2006–2008 for *Triturus cristatus* (Pabijan 2010) and involved establishing 14 study areas designated as monitoring units across the country. From 15 to 40 ponds per monitoring unit were surveyed for the presence/absence of this species, for a total of 433 surveyed ponds. Aquatic and terrestrial habitat was assessed according to the habitat suitability index for crested newts (Oldham *et al.* 2000). The method assumes that changes in local distribution can be assessed by surveying the same ponds every three to six years and counting the numbers of extirpations and colonizations (relative to previous years) for a monitoring unit. The country-wide status of this species can then be based on an appraisal of trends over all monitoring units. The methodology developed for monitoring crested newts was later adopted for 12 other species (Table 56.1) (Makomaska-Juchiewicz and Baran 2012) and modified by randomly drawing a larger number of monitoring units (to a planned sample size of >100) and also by decreasing their size (for all ponds present in 1 km^2), a method similar to that implemented in the Netherlands (Goverse *et al.* 2006). Provisions were made for rare species (*Epidalea calamita* and

Rana dalmatina) by concentrating on known or potential breeding sites. For mountain-dwelling species (*Lissotriton montandoni* and *Bombina variegata*), monitoring of transects along fixed routes combined with surveys of known breeding sites was applied.

Both monitoring methods have drawbacks. The first (Pabijan 2010) is labour-intensive at the level of the monitoring unit (because many ponds are surveyed), requiring a large investment of time from field workers. Also, this method assumes that regional amphibian demes (i.e. individuals in a single pond) form metapopulations, which may not always be the case (Smith and Green 2005). The second method (Makomaska-Juchiewicz and Baran 2012), by limiting the size of the monitoring unit to 1 km^2, ignores regional dynamics altogether and requires a large number of units (probably >100) in order to draw firm conclusions on the country-wide status of particular species, especially in the case of rare species. However, probably the greatest drawback to both methods is their reliance on commissioned experts for field surveys, thereby imposing quite severe budgetary and personnel limits. Two rounds of surveys have now been completed for 11 species at 5–8 year intervals (Table 56.1), showing declines in all but *Pelophylax* kl. *esculentus, P. ridibundus* and *Bufotes viridis* (Chief Inspectorate of Environmental Protection, 2018). However, these results should be cautiously interpreted because of the low number of surveys (3–4) at breeding sites for some species, which in consequence could have overestimated the number of extirpations for inconspicuous species or populations at low densities. Main targets for the future include integration of both methods, improvement in coverage and simplification of the assessment and reporting of habitats. A network of trained citizen scientists, NGOs and volunteers would also be a step in the right direction. More information, along with results for the first years of the monitoring programmes, can be found at http://siedliska.gios.gov.pl.

V. Conservation initiatives

Until recently, all amphibian species in Poland were protected by law (J. of Laws of 2001, No. 130, item 1456). A revision in 2016 (J. of Laws of 2016, item 2183, see Table 56.1) afforded full protection to ten species, and partial protection to eight species (the cryptic *Hyla orientalis* was not mentioned). Partial protection allows removal, harvesting or eradication under some circumstances but requires permission from Regional Directorates of Environmental Protection. Only three species were placed in the Polish Red Data Book (Głowaciński 2001): *Triturus cristatus* and *Rana dalmatina* both with near-threatened status, and *Lissotriton montandoni* of least concern.

The first attempts at amphibian conservation in Poland took place in the mid-1990s and were centred on the inspiring work of several enthusiastic herpetologists and conservationists, including Grzegorz Tabasz, Mariusz Rybacki, Marek Sołtysiak, Lars Briggs, Krzysztof Baldy, Leszek Berger and others. Greenworks, a local conservation association in southern Poland (Nowy Sącz) spearheaded amphibian conservation by implementing measures aimed at protecting breeding habitat and migrating individuals, and perhaps most importantly, by educating children and adults and procuring their involvement in local environmental protection (Rafiński and Tabasz 2001). This successful programme was soon followed by similar projects in other parts of the country. Activities included the construction of new breeding habitat and revitalization of existing ponds, protecting migration routes by installing drift fences along, and tunnels under, motorways, as well as instigating public outreach and education. These early efforts were summarized by Rybacki and Maciantowicz (2006).

The fledgling amphibian conservation movement of the 1990s soon expanded into many local initiatives too numerous to mention here. Many were presented at the Herpetological Conferences organized by the Pedagogical University in Kraków since 1990 and more recently (since 2012) at students' Herpetological Conferences at the University of Wrocław (now superseded by the

Polish Herpetological Symposium). Most national parks and many landscape parks have implemented inventory and monitoring programs and practical measures, including restoration of aquatic and terrestrial habitat, protection of breeding migration routes, workshops and education of local communities (e.g. Adaros *et al.* 2000). Ongoing projects aimed at restoring amphibian breeding habitat have succeeded in increasing species diversity and improving breeding status at some sites (Klimaszewski *et al.* 2016; Deoniziak *et al.* 2017). Many schoolteachers organize amphibian conservation-related activities for schoolchildren; however, their crucial role is usually underestimated and few have gained widespread attention (Cempulik *et al.* 2004; Olszowska 2012; Kolenda 2014). The itineraries of an increasing number of NGOs focus on amphibian conservation and public education; Table 56.2 provides links to some of these.

Some large-scale conservation projects have obtained funds from the financial instrument LIFE, in cooperation with Polish partner institutions and NGOs. For example, one project was devoted to conservation of amphibians (*Triturus cristatus* and *Bombina bombina*) and turtles (*Emys orbicularis*) (http://www.kp.org.pl/life_zolw/en/) by revitalising ponds and wetlands, by inventories and monitoring, as well as by public outreach. More recently, a second initiative in northeastern Poland (http://www.czlowiekiprzyroda.eu/life/life.htm) aims at restoring aquatic habitat and mitigating the effects of road infrastructure on amphibian populations. The Polish edition of an ongoing international project EPMAC (Educative and Participative Monitoring for wider and more effective Amphibian Conservation) was launched in 2013 by the NGOs Natura Cerca, Soontiens Ecology and the Białowieża Biodiversity Academy. This project is focused on implementing long-term monitoring of populations in northeastern Poland: in 2014, a total of 181 amphibian breeding sites were investigated (more information can be found at http://www.naturacerca.es/epmac-poland2.html).

Since 1998, the State Forests have undertaken measures to enhance the water-storage potential of Polish forests. These activities include the restoration of floodplains and wetlands by damming drainage ditches and creating spillways, re-meandering forest streams, building small reservoirs on streams and ditches, and by constructing new ponds. A total of 1124 reservoirs and ponds were built during the period 1998–2005, with an average volume slightly less than 10,000 litres. Fuelled by this success, the State Forests implemented two further water-retention projects between 2007 and 2014. The unprecedented spatial scale encompassed the entire country; one project focused on lowlands, the other on montane and highland forest. Over 6,500 small ponds, dams, dikes and

Table 56.2   A list of NGOs involved in amphibian conservation in Poland.

Name	Link
Bioróżnorodność w naszym otoczeniu	http://e-bioroznorodnosc.pl
EPMAC	http://www.naturacerca.es/epmac-poland2.html
Górnośląskie Towarzystwo Przyrodnicze	http://www.gtp.ffp.org.pl
Grupa Traszka	http://www.traszka.com.pl
Klub Gaja	http://www.klubgaja.pl/zwierzeta/plazy/
Klub Przyrodników	http://www.kp.org.pl
Nie bój żaby!	http://zaba.zrodla.org
Stowarzyszenie Człowiek i Przyroda	http://czlowiekiprzyroda.eu
Stowarzyszenie Greenworks	http://www.rytro.pl/pl/2895/0/Greenworks.html
Towarzystwo Badań i Ochrony Przyrody	http://tbop.org.pl
Towarzystwo Herpetologiczne Natrix	http://www.natrix.org.pl
Towarzystwo Ochrony Herpetofauny Tryton	http://www.tryton.waw.pl
Towarzystwo Przyrodnicze Bocian	http://bocian.org.pl/programy/plazy

meanders were constructed in less than seven years. Klimaszewski and Białaś (2013) reported stable breeding populations of amphibians in restored freshwater reservoirs in central Poland, while one of the authors (MP) has examined over a dozen of these constructions in montane habitat in southern Poland and in each case found them to constitute excellent amphibian breeding habitat. Although not specifically designed to conserve amphibians, the water-retention activities of the State Forests are nonetheless a valuable endeavour, especially since many reservoirs are built in areas with high-quality terrestrial habitat. More information can be found at http://www.ckps.lasy.gov.pl.

Importantly, a number of handbooks (in Polish) targeting amphibian conservation have appeared in print (Rafiński and Tabasz 2001; Baldy 2003; Rybacki and Maciantowicz 2006; Berger 2008; Krzysztofiak and Krzysztofiak 2016), including a very detailed guide to protecting amphibians and their habitats in the face of the expanding road infrastructure (Kurek *et al.* 2011).

VI. Conclusions

All amphibian species inhabiting Poland have large distributions that ensure their survival, at least locally. Likewise, most species (with the possible exception of *Rana dalmatina*) still have numerous populations within the boundaries of Poland, making extirpation at the national level improbable. However, it is also clear that amphibians have declined recently, and populations will continue to disappear in the near future. The cumulative effects of small-scale loss of habitat caused by drainage of wetlands and ponds, development of infrastructure, urbanization, and the degradation of habitat due to the use of agrochemicals and to disposal of refuse are the most prominent causes of amphibian declines in Poland. Recently, more weight has been placed on connectivity of populations in the landscape, and specifically the issue of roads and infrastructure as barriers to migration and their contribution to amphibians' mortality. Relatively little emphasis has been placed on the potential roles of infectious diseases, climatic change or the influence of invasive species on amphibian populations in Poland. Although attitudes towards amphibians have become more positive in recent decades, and numerous organizations and individuals have devoted their time to conservation-oriented goals, protection of amphibians in Poland is still of low priority despite legal obligations to the contrary.

VII. References

Adaros L. Ch., Briggs, L., Buszko, M., Galicki, P., Kurzawa, M. and Pabian, O., 2000. Toad talk. *The Herpetological Bulletin* **2**: 1–16.

Arciszewski, M., 2015. Migration of amphibians and their mortality on the road of Knyszyn Forest Landscape Park. *Zeszyty Naukowe Uniwersytetu Szczecińskiego, Acta Biologica* **22**: 5–14.

Babik, W. and Rafinski, J., 2001. Amphibian breeding site characteristics in the Western Carpathians, Poland. *Herpetological Journal* **11**: 41–51.

Babik, W., Szymura, J.M. and Rafiński, J., 2003. Nuclear markers, mitochondrial DNA and male secondary sexual traits variation in a newt hybrid zone (*Triturus vulgaris* × *T. montandoni*). *Molecular Ecology* **12**: 1913–30.

Babik, W., Branicki, W., Sandera, M., Litvinchuk, S., Borkin, L.J., Irwin, J.T. and Rafiński, J., 2004. Mitochondrial phylogeography of the moor frog, Rana arvalis. *Molecular Ecology* **13**: 1469–80.

Babik, W., Branicki, W., Crnobrnja-Isailović, J., Cogălniceanu, D., Sas, I., Olgun, K., Poyarkov, N.A., Garcia-París, M. and Arntzen, J.W., 2005. Phylogeography of two European newt species – discordance between mtDNA and morphology. *Molecular Ecology* **14**: 2475–91.

Babik, W., Pabijan, M., Arntzen, J.W., Cogălniceanu, D., Durka, W., and Radwan, J., 2009. Long-term survival of a urodele amphibian despite depleted major histocompatibility complex variation. *Molecular Ecology* **18**: 769–81.

Baldy, K., 2002. Płazy Gór Stołowych i ich ochrona w latach 1998–2001. *Przegląd Przyrodniczy* **13**: 63–76.

Baldy, K., 2003. *Instrukcja czynnej ochrony płazów*. Park Narodowy Gór Stołowych, Kudowa Zdrój.

Bartoszewicz, M., 1997. Śmiertelność kręgowców na szosie graniczącej z rezerwatem przyrody Słońsk. *Parki Narodowe i Rezerwaty Przyrody* **16**: 59–69.

Berger L., 1987. Impact of agriculture intensification on Amphibia. Pp. 29–32 in *Proceedings* of the 4th Ordinary General Meeting of the Societas Europaea Herpetologica, ed. by J.J. Gelder, H. Strijbosch and P.J.M. Bergers. Nijmegen.

Berger, L., 1989. Disappearance of amphibian larvae in the agricultural landscape. *Ecology International Bulletin* **17**: 65–73.

Berger, L., 2000. *Płazy i gady Polski: klucz do oznaczania*. Wydawnictwo Naukowe PWN, Warszawa.

Berger, L., 2008. *European Green frogs and their protection*. Ecological Library Foundation, Poznań.

Berger, L., Jaskowska, J. and Młynarski, M., 1969. *Płazy i gady. Katalog Fauny Polski. cz 39*. Państwowe Wydawnictwo Naukowe, Warszawa.

Błażuk, J., 2010. Śmiertelność płazów na drogach trójmiejskiego parku krajobrazowego i w jego otoczeniu. *Słupskie Prace Biologiczne* **7**: 29–50.

Bonk, M. and Pabijan, M., 2010. Changes in a regional batrachofauna in south-central Poland over a 25 year period. *North-Western Journal of Zoology* **6**: 225–44.

Bonk, M., Bury, S., Hofman, S., Szymura, J.M. and Pabijan, M., 2012. A reassessment of the northeastern distribution of *Rana dalmatina* (Bonaparte, 1840). *Herpetology Notes* **5**: 345–54.

Brzeziński, M., Eliava, G. and Żmihorski, M., 2012. Road mortality of pond-breeding amphibians during spring migrations in the Mazurian Lakeland, NE Poland. *European Journal of Wildlife Research* **58**: 685–93.

Budzik, K.A. and Budzik, K.M., 2014. A preliminary report of amphibian mortality patterns on railways. *Acta Herpetologica* **9**: 103–7.

Budzik, K.A., Budzik, K.M. and Żuwała, K., 2013. Amphibian situation in urban environment–history of the common toad *Bufo bufo* in Kraków (Poland). *Ecological Questions* **18**: 73–7.

Budzik, K.A., Budzik, K.M., Kukiełka, P., Łaptaś, A. and Bres, E.E., 2014. Water quality of urban water bodies – a threat for amphibians? *Ecological Questions* **19**: 57–65.

Chief Inspectorate of Environmental Protection, 2018. Results of The State Environmental

Monitoring Programme for the years 2015–18, unpublished.

Ciepielowski, A. and Gutry-Korycka, M., 1993. Wpływ melioracji wodnych. Pp. 313–28 in *Przemiany stosunków wodnych w Polsce w wyniku procesów naturalnych i antropogenicznych*, ed. by I. Dynowska. Uniwersytet Jagielloński, Kraków.

Civiš, P., Vojar J., Literák, I. and Balaž, V., 2012. Current state of Bd's occurrence in the Czech Republic. *Herpetological Review* **43**: 75–8.

Czarniawski, W., Gosik, R., Różycki, A. and Sałapa D., 2014. *Amphibians of Poleski National Park*. Wydawnictwo Mantis, Olsztyn.

Deoniziak, K., Hermaniuk, A. and Wereszczuk, A., 2017. Effects of wetland restoration on the amphibian community in the Narew River Valley (Northeast Poland). *Salamandra* **53**: 50–8.

Dufresnes, C., Majtyka, T., Baird, S.J., Gerchen, J.F., Borzée, A., Savary, R., Ogielska, M., Perrin, N. and Stöck, M., 2016. Empirical evidence for large X-effects in animals with undifferentiated sex chromosomes. *Scientific Reports* **6**: 21029.

Elzanowski, A., Ciesiołkiewicz, J., Kaczor, M., Radwańska, J. and Urban, R., 2009. Amphibian road mortality in Europe: a meta-analysis with new data from Poland. *European Journal of Wildlife Research* **55**: 33–43.

Fijarczyk, A., Nadachowska, K., Hofman, S., Litvinchuk, S.N., Babik, W., Stuglik, M., Gollmann, G., Choleva, L., Cogălniceanu, D., Vukov, T., Džukić, G. and Szymura, J.M., 2011. Nuclear and mitochondrial phylogeography of the European fire-bellied toads *Bombina bombina* and *Bombina variegata* supports their independent histories. *Molecular Ecology* **20**: 3381–3398.

Franz, K.W., Romanowski, J., Johst, K. and Grimm, V., 2013. Ranking landscape development scenarios affecting natterjack toad (*Bufo calamita*) population dynamics in Central Poland. *PLoS ONE* **8**: e64852.

Goverse, E., Smit, G.F., Zuiderwijk, A. and van der Meij, T., 2006. The national amphibian monitoring program in the Netherlands and NATURA 2000. Pp. 39–42 in *Proceedings of the 13th Congress of the Societas Europaea Herpetologica*, ed. by M. Vences, J. Köhler, T. Ziegler and W. Böhme. Herpetologia Bonnensis II. Bonn.

Głowaciński, Z., 2001. *Polska czerwona księga zwierząt. Kręgowce*. Państwowe Wydawnictwo Rolnicze i Leśne, Warszawa.

Głowaciński, Z. and Rafiński, J., 2003. *Atlas of the Amphibians and Reptiles of Poland. Status-Distribution-Conservation*. Biblioteka Monitoringu Środowiska, Warszawa-Kraków.

Głowaciński, Z. and Sura, P., 2018. *Atlas of the Amphibians and Reptiles of Poland. Status-Distribution-Conservation*. Wydawnictwo Naukowe PWN, Warszawa.

Głowaciński, Z. and Witkowski, Z., 1970. Numbers and biomass of amphibians estimated by the capture and removal method. *Wiadomości Ekologiczne* **16**: 328–40.

Gryz, J. and Krauze, D., 2008. Mortality of vertebrates on a road crossing the Biebrza Valley (NE Poland). *European Journal of Wildlife Research* **54**: 709–14.

Guziur, J., 2004. *Chów ryb w małych stawach*. Oficyna Wydawnicza Hoża, Warszawa.

Hetmański, T., Olech, K. and Salamon, S., 2007. Śmiertelność żaby trawnej *Rana temporaria* i ropuchy szarej *Bufo bufo* w okresie rozrodu na drodze w południowej części miasta Słupska. *Słupskie Prace Biologiczne* **4**: 15–20.

Hofman, S., Spolsky, C., Uzzell, T., Cogălniceanu, D., Babik, W. and Szymura, J.M., 2007. Phylogeography of the fire-bellied toads *Bombina*: independent Pleistocene histories inferred from mitochondrial genomes. *Molecular Ecology* **16**: 2301–16.

Juszczyk, W., 1987. *Płazy i gady krajowe*. Państwowe Wydawnictwo Naukowe, Warszawa.

Juszczyk, W., Zakrzewski, M., Zamachowski, W. and Zyśk, A., 1988. Amphibians and reptiles in the Nida Basin. *Studia Ośrodka Dokumentacji Fizjograficznej* **16**: 93–111.

Juszczyk, W., Zakrzewski, M., Zamachowski, W. and Zyśk, A., 1989. Amphibians and reptiles in the Vistula valley between

Oświęcim and Sandomierz. *Studia Ośrodka Dokumentacji Fizjograficznej* **17**: 293–306.

Kaczmarek, J.M., Kaczmarski, M. and Pędziwiatr, K., 2014. Changes in the batrachofauna in the city of Poznań over 20 years. Pp. 169–77 in *Urban fauna. Animal, Man, and the City – Interactions and Relationships*, ed. by P. Indykiewicz and J. Böhner. Bydgoszcz.

Kaczmarski, M. and Kaczmarek, J.M., 2016. Heavy traffic, low mortality-tram tracks as terrestrial habitat of newts. *Acta Herpetologica* **11**: 227–31.

Kaczmarski, M., Kolenda, K., Rozenblut-Kościsty, B. and Sośnicka, W., 2016. Phalangeal bone anomalies in the European common toad *Bufo bufo* from polluted environments. *Environmental Science and Pollution Research* **23**: 21940–6.

Klimaszewski, K. and Białaś, A., 2013. Rola zbiorników małej retencji w ochronie płazów na przykładzie Nadleśnictwa Przasnysz. *Studia i Materiały Centrum Edukacji Przyrodniczo-Leśnej* **15**: 158–64.

Klimaszewski, K., Pacholik, E. and Snopek, A., 2016. Can we enhance amphibians' habitat restoration in the post-mining areas? *Environmental Science and Pollution Research* **23**: 16941–5.

Kloskowski, J., 2010. Fish farms as amphibian habitats: factors affecting amphibian species richness and community structure at carp ponds in Poland. *Environmental Conservation* **37**: 187–94.

Kolenda, K., 2014. Active amphibian habitat protection in the South Wielkopolska Region (Poland). *Froglog* **22**: 38–9.

Kolenda, K., Senze, M. and Kowalska-Góralska, M., 2014. Zanieczyszczenia wybranych siedlisk płazów. *Chrońmy Przyrodę Ojczystą* **70**: 437–44.

Kolenda, K., Najbar, A., Ogielska, M. and Baláž, V., 2017a. *Batrachochytrium dendrobatidis* is present in Poland and associated with reduced fitness in wild populations of *Pelophylax lessonae*. *Diseases of Aquatic Organisms* **124**: 241–5.

Kolenda, K., Pietras-Lebioda, A., Hofman, S., Ogielska, M. and Pabijan, M., 2017b. Preliminary genetic data suggest the occurrence of the Balkan water frog, *Pelophylax kurtmuelleri*, in southwestern Poland. *Amphibia-Reptilia* **38**: 187–96.

Konowalik, A., Najbar, A., Babik, W., Steinfartz, S. and Ogielska, M., 2016. Genetic structure of the fire salamander *Salamandra salamandra* in the Polish Sudetes. *Amphibia-Reptilia* **37**: 405–15.

Krzysztofiak, L. and Krzysztofiak, A., 2016. *Czynna Ochrona Płazów*. Stowarzyszenie Człowiek i Przyroda, Krzywe.

Kurek, R., Rybacki, M. and Sołtysiak, M. 2011. *Poradnik ochrony płazów. Ochrona dziko żyjących zwierząt w projektowaniu inwestycji drogowych. Problemy i dobre praktyki*. Stowarzyszenie Pracownia na Rzecz Wszystkich Istot, Bystra.

Litvinchuk, S.N., Crottini, A., Federici, S., De Pous, P., Donaire, D., Andreone, F., Kalezić, M.L., Džukić, G., Lada, G.A., Borkin, L.J. and Rosanov, J.M., 2013. Phylogeographic patterns of genetic diversity in the common spadefoot toad, *Pelobates fuscus* (Anura: Pelobatidae), reveals evolutionary history, postglacial range expansion and secondary contact. *Organisms Diversity & Evolution* **13**: 433–51.

Löfvenhaft, K., Runborg, S. and Sjögren-Gulve, P., 2004. Biotope patterns and amphibian distribution as assessment tools in urban landscape planning. *Landscape and Urban Planning* **68**: 403–27.

Łonkiewicz, B. 1996. *Ochrona i zrównoważone użytkowanie lasów w Polsce*. Fundacja IUCN Poland, Warszawa.

Makomaska-Juchiewicz, M. and Baran, P. 2012. *Monitoring gatunków zwierząt. Przewodnik metodyczny. Część III*. Główny Inspektorat Ochrony Środowiska, Warszawa.

Mazgajska, J. and Mazgajski, T., 2010. Amphibians of Poland's urban areas. Preface. *Fragmenta Faunistica* **53**: 117–25.

Mioduszewski, W., 1997. Mała retencja i polityka melioracyjna. Pp. 49–62. in *Użytkowanie a ochrona zasobów wód powierzchniowych w*

Polsce, Zeszyty Naukowe Komitetu "Człowiek i Środowisko", 17. Państwowa Akademia Nauk, Warszawa.

Młynarski, M., 1987. Problemy ochrony płazów i gadów w Polsce. *Chrońmy Przyrodę Ojczystą* **3**: 18–26.

Najbar, B., Szuszkiewicz E. and Pietruszka, T., 2005. Płazy Zielonej Góry i zanikanie ich siedlisk w granicach administracyjnych miasta w latach 1974–2004. *Przegląd Zoologiczny* **49**: 155–66.

Najbar, B., Najbar, A., Maruchniak-Pasiuk, M. and Szuszkiewicz, E., 2006. Śmiertelność płazów na odcinku drogi w rejonie Zielonej Góry w latach 2003–4. *Chrońmy Przyrodę Ojczystą* **62**: 64–71.

Najbar, A., Babik, W., Najbar, B. and Ogielska, M. 2015. Genetic structure and differentiation of the fire salamander *Salamandra salamandra* at the northern margin of its range in the Carpathians. *Amphibia-Reptilia* **36**: 301–11.

Najbar, A., Rusek, A. and Najbar, B., 2017. Zagrożenia i propozycje ochronne salamandry plamistej *Salamandra salamandra* w zurbanizowanym siedlisku w Bielsku-Białej. Chrońmy Przyrodę Ojczystą **73**: 249–56.

Nawara, Z., Sendecki, P., Smolnicki, K., Szykasiuk, M., Jezierski, R. and Gąsiorowski, M., 2005. *Inwentaryzacja starorzeczy, nieużytków wodnych, oczek i zbiorników wodnych na terenie Wrocławia.* Dolnośląska Fundacja Ekorozwoju, Wrocław.

Ogielska, M. and Kierzkowski, P., 2010. Taxonomy, genetics and conservation of Amphibians in Central Europe. In *Biological diversity and nature conservation: theory and practice for teaching,* ed. by I.F. Spellerberg, M. Muhlenberg, Y. Dgebuadze and J. Slowik. Georg August University, Goettingen.

Ogielska, M., Baldy, K., Kierzkowski, P. and Maślak, R., 2008. Płazy Parku Narodowego Gór Stołowych. Pp. 237–44 in *Przyroda Parku Narodowego Gór Stołowych,* ed. by A. Witkowski, B.M. Pokryszko and W. Ciężkowski. Wydawnictwo Parku Narodowego Gór Stołowych, Kudowa Zdrój.

Ogrodowczyk, A., Ogielska, M., Kierzkowski, P. and Maślak, R., 2010. Występowanie salamandry plamistej *Salamandra s. salamandra* Linnaeus, 1758 na Dolnym Śląsku. *Przyroda Sudetów* **13**: 179–92.

Ohst, T., Gräser, Y. and Plötner, J., 2013. *Batrachochytrium dendrobatidis* in Germany: distribution, prevalences, and prediction of high risk areas. *Diseases of Aquatic Organisms* **107**: 49–59.

Olaczek, R., Kucharski, L. and Pisarek W., 1990. Zanikanie obszarów podmokłych i jego skutki środowiskowe na przykładzie województwa piotrowskiego (zlewnia Pilicy i Warty). *Studia Ośrodka Dokumentacji Fizjograficznej* **18**: 141–99.

Oldham, R.S., Keeble, J., Swan, M.J.S. and Jeffcote, M., 2000. Evaluating the suitability of habitat for the great crested newt (*Triturus cristatus*). *Herpetological Journal* **10**: 143–55.

Olszowska, M., 2012. Sześć lat realizacji projektu "Pomóżmy płazom". Pp. 76–9 in *Biologia płazów i gadów: ochrona herpetofauny,* ed. by W. Zamachowski. XI Ogólnopolska Konferencja Herpetologiczna. Wydawnictwo Naukowe Uniwersytetu Pedagogicznego, Kraków.

Orłowski, G., 2007. Spatial distribution and seasonal pattern in road mortality of the common toad *Bufo bufo* in an agricultural landscape of south-western Poland. *Amphibia-Reptilia* **28**: 25–31.

Orłowski, G., Ciesiolkiewicz, J., Kaczor, M., Radwanska, J. and Żywicka, A., 2008. Species composition and habitat correlates of amphibian roadkills in different landscapes of south-western Poland. *Polish Journal of Ecology* **56**: 659–71.

Pabijan, M., 2010. Traszka grzebieniasta *Triturus cristatus.* Pp. 32–58, in *Monitoring gatunków zwierząt. Przewodnik metodyczny. Część I,* ed. by M. Makomaska-Juchiewicz. Główny Inspektorat Ochrony Środowiska, Warszawa.

Pabijan, M. and Babik, W., 2006. Genetic structure in northeastern populations of the Alpine newt (*Triturus alpestris*): evidence for post-Pleistocene differentiation. *Molecular Ecology* **15**: 2397–407.

Pabijan, M., Babik, W. and Rafiński, J., 2005. Conservation units in north-eastern populations of the Alpine newt (*Triturus alpestris*). *Conservation Genetics* **6**: 307–12.

Pabijan, M., Rożej, E. and Bonk, M., 2009. An isolated locality of the alpine newt (*Mesotriton alpestris* Laurenti, 1768) in central Poland. *Herpetology Notes* **2**: 23–6.

Pabijan, M., Zieliński, P., Dudek, K., Chloupek, M., Sotiropoulos, K., Liana, M. and Babik, W., 2015. The dissection of a Pleistocene refugium: phylogeography of the smooth newt, *Lissotriton vulgaris*, in the Balkans. *Journal of Biogeography* **42**: 671–83.

Pabijan, M., Zieliński, P., Dudek, K., Stuglik, M. and Babik, W., 2017. Isolation and gene flow in a speciation continuum in newts. *Molecular Phylogenetics and Evolution* **116**: 1–12.

Rafiński, J. and Tabasz, G., 2001. *Ochrona płazów*. Greenworks, Nowy Sącz.

Recuero, E., Buckley, D., García-París, M., Arntzen, J.W., Cogălniceanu, D. and Martínez-Solano, I., 2014. Evolutionary history of *Ichthyosaura alpestris* (Caudata, Salamandridae) inferred from the combined analysis of nuclear and mitochondrial markers. *Molecular Phylogenetics and Evolution* **81**: 207–20.

Reshetnikov, A.N., 2003. The introduced fish, rotan (*Perccottus glenii*), depresses populations of aquatic animals (macroinvertebrates, amphibians, and a fish). *Hydrobiologia* **510**: 83–90.

Rybacki, M., 1995. Zagrożenie płazów na drogach Pienińskiego Parku Narodowego. *Pieniny. Przyroda i Człowiek* **4**: 85–97.

Rybacki, M. and Krupa, A., 2002. Wstępny raport na temat śmiertelności płazów na drogach parków krajobrazowych Województwa Wielkopolskiego. *Przegląd Przyrodniczy* **13**: 77–86.

Rybacki, M. and Berger, L. 2003. Współczesna fauna płazów Wielkopolski na tle zaniku ich siedlisk rozrodczych. Pp. 143–73 in *Stepowienie Wielkopolski – pół wieku później*, ed. by J. Banaszak. Wydawnictwo Akademii Bydgoskiej. Bydgoszcz.

Rybacki, M. and Domańska, E., 2004. Intensywność migracji i śmiertelność płazów na drogach gospodarstwa rybackiego Oleśnica (powiat Chodzież, województwo wielkopolskie). Pp. 90–4 in *Biologia płazów i gadów: ochrona herpetofauny*, ed. by W. Zamachowski. VII Ogólnopolska Konferencja Herpetologiczna. Wydawnictwo Naukowe Uniwersytetu Pedagogicznego, Kraków.

Rybacki, M. and Maciantowicz, M., 2006. *Ochrona żółwia błotnego, traszki grzebieniastej i kumaka nizinnego*. Wydawnictwo Klubu Przyrodników, Świebodzin.

Sadza, I., Oleś W., Zając, B., Bury, S., Żuwała, K. and Pabijan, M., 2015. Aktualne rozmieszczenie płazów na terenie Tatrzańskiego Parku Narodowego na tle badań prowadzonych w drugiej połowie XX wieku. Pp. 47–54 in *Przyroda Tatrzańskiego Parku Narodowego a Człowiek. Tom II Nauki Biologiczne*. Wydawnictwa Tatrzańskiego Parku Narodowego, Zakopane.

Schimscheiner, L., 1990. Zmiany liczebnosci kumaka nizinnego *Bombina bombina* L. w stawach rybnych w Mydlnikach. Pp. 13–4 in *Streszczenia referatów, II Ogólnopolska Konferencja Herpetologiczna*, ed. by W. Zamachowski. Wydawnictwo Naukowe Uniwersytetu Pedagogicznego, Kraków.

Semlitsch, R.D., 2003. *Amphibian Conservation*. Smithsonian Books, Washington and London.

Smith, M.A. and Green, D.M., 2005. Dispersal and the metapopulation paradigm in amphibian ecology and conservation: are all amphibian populations metapopulations? *Ecography* **28**: 110–28.

Sołtysiak, M., 2006. Ocena przyszłego wpływu budowanej aktualnie obwodnicy Grodźca Śląskiego na populacje płazów. Pp. 144–8 in *Biologia płazów i gadów: ochrona herpetofauny*, ed. by W. Zamachowski. VIII Ogólnopolska Konferencja Herpetologiczna. Wydawnictwo Naukowe Uniwersytetu Pedagogicznego, Kraków.

Sołtysiak, M. and Matusiak, R. 2006. Ochrona herpetofauny w pasie budowy Drogowej Trasy Średnicowej w Rudzie Śląskiej. Pp.

149–53 in *Biologia płazów i gadów: ochrona herpetofauny*, ed. by W. Zamachowski. VIII Ogólnopolska Konferencja Herpetologiczna. Wydawnictwo Naukowe Uniwersytetu Pedagogicznego, Kraków.

Sołtysiak, M. and Kaźmierczak, J. 2008. Weryfikacja raportu oceny oddziaływania na środowisko autostrady A1 odcinka Sośnica-Bełk w aspekcie oddziaływania inwestycji na płazy. Pp. 125–30 in *Biologia płazów i gadów: ochrona herpetofauny*, ed. by W. Zamachowski. IX Ogólnopolska Konferencja Herpetologiczna. Wydawnictwo Naukowe Uniwersytetu Pedagogicznego, Kraków.

Sołtysiak M. and Rybacki, M. 2010. Złe praktyki w opracowywaniu i opiniowaniu raportów oceny oddziaływania na środowisko w zakresie herpetologii przy inwestycjach drogowych na przykładzie obwodnicy Grodźca Ślaskiego (województwo śląskie). Pp. 150–6 in *Biologia płazów i gadów: ochrona herpetofauny*, ed. by W. Zamachowski. X Ogólnopolska Konferencja Herpetologiczna. Wydawnictwo Naukowe Uniwersytetu Pedagogicznego, Kraków.

Sołtysiak M. and Rybacki, M. 2012. Wpływ jakości raportów środowiskowych na efektywne planowanie kompenscji przyrodniczej dla płazów na przykładzie projektowanego odcinka G1 Drogowej Trasy Średnicowej (DTŚ). Pp. 125–30 in *Biologia płazów i gadów: ochrona herpetofauny*, ed. by W. Zamachowski. XI Ogólnopolska Konferencja Herpetologiczna. Wydawnictwo Naukowe Uniwersytetu Pedagogicznego, Kraków.

Sotiropoulos, K., Eleftherakos, K., Džukić, G., Kalezić, M.L., Legakis, A. and Polymeni, R.M., 2007. Phylogeny and biogeography of the alpine newt *Mesotriton alpestris* (Salamandridae, Caudata), inferred from mtDNA sequences. *Molecular Phylogenetics and Evolution* 45: 211–26.

Speybroeck, J., Beukema, W. and Crochet, P.A., 2010. A tentative species list of the European herpetofauna (Amphibia and Reptilia) – an update. *Zootaxa* 2492: 1–27.

Spitzen-van der Sluijs, A., Martel, A., Asselberghs, J., Bales, E.K., Beukema, W., Bletz, M.C., Dalbeck, L., Goverse, E., Kerres, A., Kinet, T., Kirst, K., Laudelout, A., Marin da Fonte, L.F., Nöllert, A., Ohlhoff, D., Sabino-Pinto, J., Schmidt, B.R., Speybroeck, J., Spikmans, F., Steinfarz, S., Veith, M., Vences, V., Wagner, N., Pasmans, F. and Lötters, S., 2016. Expanding distribution of lethal amphibian fungus *Batrachochytrium salamandrivorans* in Europe. *Emerging Infectious Diseases*, 22: 1286–8.

Stöck, M., Moritz, C., Hickerson, M., Frynta, D., Dujsebayeva, T., Eremchenko, V., Macey, J.R., Papenfuss, T.J. and Wake, D.B., 2006. Evolution of mitochondrial relationships and biogeography of Palearctic green toads (*Bufo viridis* subgroup) with insights in their genomic plasticity. *Molecular Phylogenetics and Evolution* 41: 663–89.

Stöck, M., Dufresnes, C., Litvinchuk, S.N., Lymberakis, P., Biollay, S., Berroneau, M., Borzée, A., Ghali, K., Ogielska, M. and Perrin, N., 2012. Cryptic diversity among Western Palearctic tree frogs: Postglacial range expansion, range limits, and secondary contacts of three European tree frog lineages (*Hyla arborea* group). *Molecular Phylogenetics and Evolution* 65: 1–9.

Sura, P., Janulis, E. and Profus, P., 2010. Chytridiomikoza-smiertelne zagrozenie dla plazow. *Chrońmy Przyrodę Ojczystą* 66: 406–21.

Szymura, J.M., 1993. Analysis of hybrid zones with *Bombina*. Pp. 261–89 in *Hybrid Zones and the Evolutionary Process*, ed. by R.G. Harisson. Oxford University Press, New York.

Szymura, J.M., Uzzell, T. and Spolsky, C., 2000. Mitochondrial DNA variation in the hybridizing fire-bellied toads, *Bombina bombina* and *B. variegata*. *Molecular Ecology* 9: 891–9.

Szyndlar, Z., 1994/1995. Płazy i gady Ojcowskiego Parku Narodowego: stan w końcu lat osiemdziesiątych. *Prądnik, Prace i Materiały Muzeum Szafera* 9: 231–40.

Świerad, J., 2003. *Płazy i gady Tatr, Podhala, Doliny Dunajca oraz ich ochrona*. Wydawnictwo Naukowe Akademii Pedagogicznej, Kraków.

Tomalka-Sadownik, A. and Rozenblut-Kościsty, B., 2010. Amphibians of Wałbrzych. *Fragmenta Faunistica* **53**: 163–79.

Tryjanowski, P., Rybacki, M. and Sparks, T., 2003. Changes in the first spawning dates of common frogs and common toads in western Poland in 1978–2002. *Annales Zoologica Fennica* **40**: 459–64.

Tryjanowski, P., Sparks, T., Rybacki, M. and Berger, L., 2006. Is body size of the water frog *Rana esculenta* complex responding to climate change? *Naturwissenschaften* **93**: 110–3.

Wielstra, B., Crnobrnja-Isailović, J., Litvinchuk, S.N., Reijnen, B.T., Skidmore, A.K., Sotiropoulos, K., Toxopeus, A.G., Tzankov, N., Vukov, T. and Arntzen, J.W., 2013. Tracing glacial refugia of *Triturus* newts based on mitochondrial DNA phylogeography and species distribution modeling. *Frontiers in Zoology* **10**: 13.

Wirga, M. and Majtyka, T., 2015. Herpetofauna of the opencast mines in Lower Silesia (Poland). *Fragmenta Faunistica* **58**: 65–70.

Witkowski, A. 2011. *Perccottus glenii* (Dybowski, 1877). Pp. 423–428 in *Gatunki obce w faunie Polski*, ed. by Z. Głowaciński, H. Okarma, J. Pawłowski and W. Solarz. Instytut Ochrony Przyrody PAN, Kraków.

Wojdan, D., 2010. Impact of vehicle traffic on amphibian migrations in the protection zone of the Świętokrzyski National Park. *Teka Komisji Ochrony i Kształtowania Środowiska Przyrodniczego* **7**: 466–72.

Wołk, K., 1978. Zabijanie zwierząt przez pojazdy samochodowe w Rezerwacie Krajobrazowym Puszczy Białowieskiej. *Chrońmy Przyrodę Ojczystą* **6**: 20–8.

Woś, A., 1999. *Klimat Polski*. Wydawnictwo Naukowe PWN, Warszawa.

Zakrzewski, M., 2007. *Salamandra plamista. Rozmieszczenie, biologia i zagrożenia*. Wydawnictwo Naukowe Akademii Pedagogicznej, Kraków.

Zamachowski, W. and Plewa, G., 1996. Śmiertelność płazów podczas wędrówek. Pp. 89–91 in *Materiały IV Ogólnopolskiej Konferencji Herpetologicznej*, ed. by W. Zamachowski. Wydawnictwo Naukowe Uniwersytetu Pedagogicznego, Kraków.

Zieliński, P., Nadachowska-Brzyska, K., Wielstra, B., Szkotak, R., Covaciu-Marcov, S.D., Cogălniceanu, D. and Babik, W., 2013. No evidence for nuclear introgression despite complete mtDNA replacement in the Carpathian newt (*Lissotriton montandoni*). *Molecular Ecology* **22**: 1884–903.

57 Amphibian conservation in Switzerland

Benedikt R. Schmidt and Silvia Zumbach

Abbreviations or acronyms used in the text or references

BAFU	*Bundesamt für Umwelt*
BLW	*Bundesamt für Landwirtschaft*
CSCF	*Centre Suisse pour la Cartographie de la Faune*
NGO	*Non-governmental Organization*
NHG	*Natur- und Heimatschutzgesetz*

I. Introduction

Switzerland is a landlocked country in western Europe. The area is about 41,000 km² and the human population is nearly 8 million. Almost a third of Switzerland (30.3%) is covered by forest or scrubland, 38.3% is used for agricultural purposes, 5.9% is covered by built-up areas (including transportation infrastructure), 4.2% by lakes and rivers, and the remaining 21.3% by unproductive land (Schmidt and Zumbach 2008). Switzerland is divided into three main landscapes: the relatively dry Jura mountain range, the Central Plateau and the Alps mountain range (where most of the unproductive land is found) (Schmidt and Zumbach 2008). The three main landscapes cover roughly 10%, 30%, and 60% of the area respectively. Amphibian diversity is greatest in the lowlands north (the Central Plateau) and south of the Alps. Because Switzerland covers some area both north and south of the Alps, the amphibian fauna includes northern European as well as Mediterranean species and subspecies (the latter include *Lissotriton vulgaris meridionalis*, *Hyla intermedia* and *Rana latastei*).

II. Declining species of amphibians and/or species of special conservation concern

According to IUCN's global Red List of amphibians, only one native Swiss amphibian species, the Italian Agile Frog *Rana latastei*, is globally threatened. In the most recent national Red List (Schmidt and Zumbach 2005), 70% of the native species were red-listed (Table 57.1). Many species were red-listed because of dramatic population declines (i.e., extirpations of populations) between 1985 and 2005 (Table 57.1; see also Cruickshank *et al.* 2016). Large-scale and long-term declines of common species were also documented (Petrovan and Schmidt 2016). However, declines and regional extirpations of species were documented decades ago (Hotz and Broggi 1982; Grossenbacher 1988). Declines in the 1960s and 1970s may have been stronger than in later decades. The first national Red List was published in 1982 (Hotz and Broggi 1982) and listed 15 of 19 species as endangered.

Table 57.1   Red list status of Swiss amphibians, magnitude of population declines in the period 1985–2005, range in size, and reasons for red list status. The red list assessment was based on the criteria of the IUCN. Hence, there are two main reasons why species could be red-listed: either because there were population declines or because the distributional range of the species within Switzerland is small. Estimates of decline and range are from Schmidt and Zumbach (2005).

Species	Red list status	Decline (%)	Range (km²)	Reason for red-list status
Salamandra atra	LC	–	–	–
Salamandra salamandra	VU	–	–	population decline
Ichthyosaura alpestris	LC	–17.9	4,572	–
Lissotriton helveticus	VU	–32.7	1,240	population decline, small range
Lissotriton vulgaris	EN	–48.2	157	population decline, small range
Triturus cristatus	EN	–56.7	128	population decline, small range
Triturus carnifex	EN	–29.1	36	small range
Alytes obstetricans	EN	–52.5	2,059	population decline
Bombina variegata	EN	–56.4	2,795	population decline
Pelobates fuscus	DD	–	–	–
Hyla intermedia	EN	–18.1	233	small range
Hyla arborea	EN	–56.0	1,210	population decline
Rana arvalis	DD	–	–	–
Rana temporaria	LC	–14.2	24,037	–
Rana dalmatina	EN	–26.5	487	small range
Rana latastei	VU	–	–	small population size
Pelophylax lessonae/P. esculentus	NT	–21.9	7,257	population decline
Pelophylax ridibundus	NE	–	–	invasive species
Bufo bufo	VU	–31.2	12,029	population decline
Epidalea calamita	EN	–62.5	1,152	population decline
Bufotes viridis	RE	–	–	–

The proportion of red-listed species was higher in 1982 than in 2005 but the difference is mainly caused by differences in methodology and because additional species were assessed (because new species, such as *Hyla intermedia,* were described).

All amphibian species native to Switzerland also occur in other European countries. One species for which Switzerland carries some responsibility is the Alpine salamander *Salamandra atra* because a large proportion of the range of this species is within Switzerland. There are no conservation actions for this species, however, because it is still considered to be widespread and abundant. This is due to the fact that its habitats at high elevations are less affected by human activities than are amphibian habitats in the lowlands.

There are two invasive amphibian species that are of conservation concern. The frog *Pelophylax ridibundus* is a threat to native species because it is probably an important predator (Roth *et al.* 2016; Dufresnes *et al.* 2018). It also threatens the persistence of the peculiar hybridogenetic system (Vorburger and Reyer 2003). Recent research showed that the situation is more complex than previously thought because there are many cryptic invasions of *Pelophylax* frogs (Dubey *et al.* 2014). The second species is the newt *Triturus carnifex,* which is native to southern Switzerland, and which has been released in the Geneva area; preliminary evidence suggests that it was also introduced in other areas. In the Geneva area, it hybridized with and replaced the native *T. cristatus* (Arntzen and Thorpe 1999; Dufresnes *et al.* 2016).

III. Conservation measures and monitoring programmes

The loss of wetlands is the main reason for the decline of amphibians. About 90% of wetlands have been drained or otherwise destroyed (Imboden 1976; Gimmi *et al.* 2011). The loss of habitats is associated with a loss of the quality of remaining habitat (e.g. shrinking size of wetlands, isolation of wetlands, introduction of non-native fish, intensive use of the surrounding agricultural matrix including extensive drainage systems, and high density of roads (Hotz and Broggi 1982; Schmidt and Zumbach 2005, 2010).

Amphibian conservation has a long history in Switzerland (Heusser 1968; Meisterhans and Heusser 1970). A first field guide to the Swiss amphibians and their natural history and conservation was published in 1966 (Brodmann 1966).

Amphibian species and their habitats have been protected by law since 1966. Since then, it is forbidden to kill amphibians or harm them in any way. Legal protection also includes the wetlands that serve as breeding sites. A particular feature of the Swiss nature conservation law (Natur- und Heimatschutzgesetz NHG) is that all habitats where species (e.g. amphibians) occur that are protected by law or that feature on the Red Lists published by the Swiss federal agency of the environment, may only be destroyed if there is some mitigation measure that compensates for the loss. Unfortunately, legal protection did not prevent the dramatic declines that occurred in past decades. To preserve the most important amphibian breeding sites, a network of amphibian breeding sites of federal importance, which included about 800 sites, was established in 2001 (Ryser 2002). The algorithm that was used to determine which sites should be declared to be of federal importance was based on the rarity of species and population sizes (Borgula *et al.* 1994). Amphibian breeding sites of federal importance enjoy special protection because they can only be destroyed if a development project, e.g. a new highway, is also of federal importance. Nevertheless, even sites of federal importance are destroyed (but the loss mitigated); see Schmidt (2008) for a case study.

Surveys and systematic inventories of amphibians and their breeding sites have a long history in Switzerland. The first systematic survey in the canton of Zurich dates back to the late 1960s (Escher 1972). Systematic surveys were conducted in almost all cantons such that a comprehensive atlas of the distributions of the Swiss amphibians could be published in 1988 (Grossenbacher 1988). Updated distribution maps were published by Meyer *et al.* (2009).

Of particular importance for amphibian conservation was the creation of Info Fauna Karch (formerly "Karch", the acronym for "Koordinationsstelle für Amphibien- und Reptilienschutz in der Schweiz", see http://www.karch.ch), the Swiss Amphibian and Reptile Conservation Program, in 1979. Info Fauna Karch maintains the Swiss amphibian distribution database (as part of the Swiss Centre for Cartography of the Fauna CSCF). The Info Fauna Karch database currently (June 2018) holds 267,262 amphibian records, and 13,474 amphibian breeding sites are registered. The distribution data are available online (at http://lepus.unine.ch/carto/) and can be used by scientists and conservation practitioners. The data are often used during environmental impact assessments. In the future, the data will also be used to estimate temporal trends in the occurrence of amphibians, using a method similar to the one presented by Kéry *et al.* (2010). Apart from maintaining a database, Info Fauna Karch provides a free consultancy service for all aspects of amphibian and reptilian conservation. Info Fauna Karch closely collaborates with the federal agency for the environment, maintains a network of representatives in almost all cantons, and collaborates with scientists at Swiss universities and abroad; the latter also includes the joint supervision of master and doctoral students. Info Fauna Karch regularly publishes booklets and scholarly articles on the conservation of amphibians (e.g. on such topics as amphibians and roads, amphibians in sewage systems, guidelines for the conservation of endangered species). It also maintains a library that includes a large body of grey literature on the conservation of amphibians. Info Fauna Karch organizes an annual one-day conference on the ecology, conservation and natural history of amphibians and reptiles.

At the national level, there is unfortunately no systematic monitoring of amphibians. There is only a systematic monitoring of amphibians in the canton of Aargau (Meier and Schelbert 1999; Schmidt 2005). Other cantons have ongoing monitoring programs for some species and/or sites. A systematic monitoring of the amphibians inhabiting the amphibian breeding sites of federal importance was initiated in 2011 (Boch *et al.* 2018).

IV. Conclusions and summary

Amphibian conservation in Switzerland is well organized and amphibians are popular species. Hence, the preconditions for successful conservation are present. Nevertheless, the strong population declines documented by the 2005 Red List (Schmidt and Zumbach 2005) show that amphibian conservation does not function as effectively as it could or should. In our opinion, there are several main deficiencies (Schmidt and Zumbach 2010). First, there is still an ongoing loss of habitats and, equally important, a loss of quality of the habitats that remain. Second, while the conservation laws are very good and could be powerful means of protecting amphibians, the laws are often not enforced and there are still many laws in other domains that hinder amphibian conservation. For example, amphibians' habitats are not included in the list of agricultural set-asides for which farmers may receive subsidies. Hence, there is little incentive for farmers to have ponds or wetlands on their land. The process of changing agricultural policy is ongoing; at least amphibians are now included in the official list of the target species for conservation on agricultural land ("Umweltziele Landwirtschaft", BAFU and BLW 2008). Most amphibian species are also included among the ~3000 species that have priority for conservation at the national level (BAFU 2011). Nevertheless, there is unreadiness on the part of the many stakeholders to give priority to conservation. Third, there is a lack of temporary ponds (Schmidt *et al.* 2015). While digging ponds is a popular activity among conservation practitioners, most of these ponds are permanent and unsuitable for most of the threatened species. Karch has therefore started a campaign that stresses the need for more temporary ponds. This highlights the fact that conservationists often do not know how to protect amphibians most efficiently. As a consequence of the Karch campaign, Pro Natura, a major Swiss conservation NGO, now also recommends temporary ponds for amphibian conservation and has published a booklet on the construction of temporary ponds (Pellet 2014).

We conclude that despite legal protection of amphibian species and their habitats since the mid-1960s, there were, are, and will be, strong population declines. However, if one is willing to invest time, effort and money, then – as numerous examples show – amphibian conservation can be successful and amphibian populations could thrive even in a country with a high human population density and in a landscape that has been modified by humans for millennia.

V. References

Arntzen, J.W. and Thorpe, R.S., 1999. Italian crested newts (*Triturus carnifex*) in the Basin of Geneva: Distribution and genetic interactions with autochthonous species. *Herpetologica* **55**: 423–33.

BAFU, 2011. *Liste der National Prioritären Arten: Arten mit nationaler Priorität für die Erhaltung und Förderung, Stand 2010.* Umwelt-Vollzug Nr. 1103. Bundesamt für Umwelt, Bern.

BAFU and BLW, 2008. *Umweltziele Landwirtschaft: Hergeleitet aus bestehenden rechtlichen Grundlagen.* Umwelt-Wissen Nr. 0820. Bundesamt für Umwelt, Bern.

Boch, S., Ginzler, C., Schmidt, B.R., Bedolla, A., Ecker, K., Graf, U., Küchler, H., Küchler, M., Holderegger, R. and Bergamini, A., 2018. Wirkt der Schutz von Biotopen? Ein Programm zum Monitoring der Biotope von nationaler Bedeutung in der Schweiz. *Anliegen Natur* **40**: 39–48.

Borgula, A., Fallot, P. and Ryser, J., 1994. *Inventar der Amphibienlaichgebiete von nationaler Bedeutung: Schlussbericht.* Schriftenreihe Umwelt Nr. 223. Bundesamt für Umwelt, Wald und Landschaft, Bern.

Brodmann, P., 1966. Die Amphibien der Basler Region. *Veröffentlichungen aus dem Naturhistorischen Museum Basel* **4**: 1–32.

Cruickshank, S.S., Ozgul, A. Zumbach, S. and Schmidt, B.R., 2016. Quantifying population declines based on presence-only records for Red List assessments. *Conservation Biology* **30**: 1112–21.

Dubey, S., Leuenberger, J. and Perrin, N. 2014. Multiple origins of invasive and 'native' water frogs (*Pelophylax* spp.) in Switzerland. *Biological Journal of the Linnean Society* **112**: 442–9.

Dufresnes, C., Leuenberger, J., Amrhein, V., Bühler, C., Thiébaud, C., Bohnenstengel, T. and Dubey, S., 2018. Invasion genetics of marsh frogs (*Pelophylax ridibundus sensu lato*) in Switzerland. *Biological Journal of the Linnean Society* **123**: 402–10.

Dufresnes, C., Pellet, J., Bettinelli-Riccardi, S., Thiébaud, J., Perrin, N. and Fumagalli, L., 2016. Massive genetic introgression in threatened northern crested newts (*Triturus cristatus*) by an invasive congener (*T. carnifex*) in Western Switzerland. *Conservation Genetics* **17**: 839–46.

Escher, K., 1972. Die Amphibien des Kantons Zürich. *Vierteljahresschrift der naturforschenden Gesellschaft Zürich* **117**: 335–80.

Gimmi, U., Lachat, T. and Bürgi, M., 2011. Reconstructing the collapse of wetland networks in the Swiss lowlands 1850–2000. *Landscape Ecology* **26**: 1071–83.

Grossenbacher, K., 1988. Verbreitungsatlas der Amphibien der Schweiz. *Documenta faunistica helvetiae* **7**: 1–207.

Heusser, H., 1968. Wie Amphibien schützen? *Flugblatt Naturforschende Gesellschaft Schaffhausen* **3**: 1–14.

Hotz, H. and Broggi, M.F., 1982. *Rote Liste der gefährdeten und seltenen Amphibien und Reptilien der Schweiz.* Schweizerischer Bund für Naturschutz, Basel.

Imboden, C., 1976. *Leben am Wasser: Kleine Einführung in die Lebensgemeinschaften der Feuchtgebiete.* Schweizerischer Bund für Naturschutz, Basel.

Kéry, M., Royle, J.A., Schmid, H., Schaub, M., Volet, B., Häfliger, G. and Zbinden, N., 2010. Site-occupancy distribution modeling to correct population-trend estimates derived from opportunistic observations. *Conservation Biology* **24**: 1388–97.

Meier, C. and Schelbert, B., 1999. Amphibienschutzkonzept Kanton Aargau. *Aargauer Naturforschende Gesellschaft Mitteilungen* **35**: 41–69.

Meisterhans, K. and Heusser, H., 1970. Amphibien und ihre Lebensräume: Gefährdung – Forschung – Schutz. *Natur und Mensch* **12**: 1–20.

Meyer, A., Zumbach, S., Schmidt, B. and Monney, J.-C. 2009. *Auf Schlangenspuren und Krötenpfaden: Amphibien und Reptilien der Schweiz.* Haupt Verlag, Bern.

Pellet, J., 2014. *Temporäre Gewässer für gefährdete Amphibien schaffen: Leitfaden für die Praxis.* Pro Natura, Basel.

Petrovan, S.O. and Schmidt, B.R., 2016. Volunteer conservation action data reveals large-scale and long-term negative population trends of a widespread amphibian, the common toad (*Bufo bufo*). *PLoS ONE* **11**: e0161943. doi:10.1371/journal.pone.0161943.

Roth, T., Bühler, C. and Amrhein, V., 2016. Estimating effects of species interactions on populations of endangered species. *American Naturalist* **187**: 457–67.

Ryser, J., 2002. *Bundesinventar der Amphibienlaichgebiete von nationaler Bedeutung: Vollzugshilfe*. Bundesamt für Umwelt, Wald und Landschaft, Bern.

Schmidt, B.R., 2005. Monitoring the distribution of pond-breeding amphibians when species are detected imperfectly. *Aquatic Conservation* **15**: 681–92.

Schmidt, B.R., 2008. Umsiedlung einer Amphibienpopulation in der Schweiz. Pp. 112–3 in *Amphibien brauchen unsere Hilfe: Verhandlungsbericht des Amphibienkurses in Chemnitz, 27.–30. Juni 2007*, ed. by P. Dollinger. World Association of Zoos and Aquariums, Bern.

Schmidt, B.R. and Zumbach, S., 2005. *Rote Liste der gefährdeten Amphibien der Schweiz*. Bundesamt für Umwelt, Wald und Landschaft, Bern.

Schmidt, B.R. and Zumbach, S., 2008. Amphibian road mortality and how to prevent it: a review. Pp. 157–67 in *Urban Herpetology*, ed. by J.C. Mitchell, R.E. Jung Brown and B. Bartholomew. *Herpetological Conservation* **3**. Society for the Study of Amphibians and Reptiles, Salt Lake City.

Schmidt, B.R. and Zumbach, S., 2010. Neue Herausforderungen und Wege im Amphibienschutz. *Wildbiologie* **4/37**: 1–16.

Schmidt, B.R., Zumbach, S., Tobler, U. and Lippuner, M., 2015. Amphibien brauchen temporäre Gewässer. *Zeitschrift für Feldherpetologie* **22**: 137–50.

Vorburger, C. and Reyer, H.-U., 2003. A genetic mechanism of species replacement in European waterfrogs? *Conservation Genetics* **4**: 141–55.

58 Amphibian declines and conservation in Austria

Marc Sztatecsny

Abbreviations and acronyms used in the text and references

Bd	Batrachochytrium dendrobatidis
FWF	*Austrian Science Fund*
IUCN	*International Union for the Conservation of Nature*
m a.s.l.	*metres above sea level*
ÖGH	*Österreichische Gesellschaft für Herpetologie*
UMG	*Umweltbüro Grabher*

I. Climate and the amphibian fauna

Austria is a land-locked country in the centre of Europe, with 65% of its area consisting of the Alps that traverse the country from east to west. The Alps form a barrier to fronts typically advecting from the Atlantic and have a significant influence on the local climate (Auer *et al.* 2007). Western and northwestern Austria is influenced by a temperate maritime regime, in which uplifting of air along the edge of the mountains causes high mean annual precipitation (regionally >1,800 mm/year). The central parts of the Alps, as well as the eastern and southeastern lowlands, receive less precipitation, and areas of low elevation exhibit a warm continental climate (mean annual precipitation regionally < 600 mm/year). The south is influenced by a sub-Mediterranean climatic regime.

The 21 amphibian species occurring in Austria (Table 58.1) have central, eastern or southern European distributions (Gasc *et al.* 1997), except for *Bufo [Epidalea] calamita* and *Lissotriton helveticus* that are known only from two populations and a single population, respectively, close to the country's border (Cabela *et al.* 2001; Grabher and Niederer 2011). The distribution of amphibians within Austria is largely determined by elevation. The typical lowland species, predominating in assemblages below 300 m a.s.l., are *Triturus dobrogicus*, *Bombina bombina*, *Bufo [Pseudepidalea] viridis*, *Pelobates fuscus*, *Hyla arborea*, *Pelophylax esculentus*, *P. ridibundus*, *Rana arvalis* and *R. dalmatina*. Middle to high elevations (1,000–2,000 m) are dominated by *Ichthyosaura alpestris*, *R. temporaria*, *Bufo bufo* and *Salamandra atra*, with the first three species occurring from 200 m upward (the highest record for an amphibian in Austria is 2,850 m for *R. temporaria*) (Cabela *et al.* 2001).

II. Amphibian declines and their causes

As in large parts of the rest of Europe, loss and fragmentation of habitats are the most severe threats to Austria's amphibian populations (Stuart *et al.* 2004; Gollmann 2007). Because alteration of the

Table 58.1   Amphibian species occurring in Austria (names according to Frost 2011) and their IUCN status for Austria and Europe (Gollmann 2007; IUCN 2011).

Species	English Name	German Name	IUCN Red List Criteria (Austria)	IUCN Red List Criteria (Europe)
Bombina bombina (Linnaeus, 1761)	European Fire-bellied Toad	Rotbauchunke	VU	LC
Bombina variegata (Linnaeus, 1758)	Yellow-bellied Toad	Gelbbauchunke	VU	LC
Bufo bufo (Linnaeus, 1758)	Common Toad	Erdkröte	NT	LC
Bufo [Epidalea] calamita (Laurenti, 1768)	Natterjack Toad	Kreuzkröte	CR	LC
Bufo [Pseudepidalea] viridis (Laurenti, 1768)	European Green Toad	Wechselkröte	VU	LC
Hyla arborea (Linnaeus, 1758)	European Treefrog	Europäischer Laubfrosch	VU	LC
Pelobates fuscus (Laurenti, 1768)	Common Spadefoot Toad	Knoblauchkröte	EN	LC
Pelophylax esculentus (Linnaeus, 1758)	Edible Frog	Teichfrosch	NT	LC
Pelophylax lessonae (Camerano, 1882)	Pool Frog	Kleiner Wasserfrosch	VU	LC
Pelophylax ridibundus (Pallas, 1771)	Eurasian Marsh Frog	Seefrosch	VU	LC
Rana arvalis (Nilsson, 1842)	Moor Frog	Moorfrosch	VU	LC
Rana dalmatina (Fitzinger, 1839)	Agile Frog	Springfrosch	NT	LC
Rana temporaria (Linnaeus, 1758)	Common Frog	Grasfrosch	NT	LC
Ichthyosaura alpestris (Laurenti, 1768)	Alpine Newt	Bergmolch	NT	LC
Lissotriton helveticus (Razoumowsky, 1789)	Palmate Newt	Fadenmolch	DD	LC
Lissotriton vulgaris (Linnaeus, 1758)	Smooth Newt	Teichmolch	NT	LC
Salamandra atra (Laurenti, 1768)	Alpine Salamander	Alpensalamander	NT	LC
Salamandra salamandra (Linnaeus, 1758)	Common Fire Salamander	Feuersalamander	NT	LC
Triturus carnifex (Laurenti, 1768)	Italian Crested Newt	Alpenkammmolch	VU	LC
Triturus cristatus (Laurenti, 1768)	Northern Crested Newt	Nördlicher Kammmolch	EN	LC
Triturus dobrogicus (Kiritzescu, 1903)	Danube Crested Newt	Donaukammmolch	EN	NT

landscape has a long history, consulting reports from decades ago reveal what has been lost. In a treatise on European amphibians, Brehm and Schmidtlein (1902) stated:

> "No other frog species has more, no other so many enemies (predators) as the Common Frog (*Rana temporaria*) … The almost uncountable number of predators is joined by man because even more than water frogs the common frog is being captured and slaughtered for its stout legs. The thousands that lose their lives, however, do not reduce the overall number of these valuable animals or at least not noticeably. Spring conditions favourable for frog breeding may compensate for losses of 10 previous years."

Still, in 1976, a landlord from the Austrian province of Osttirol applied for permission to kill 5,000 *R. temporaria* in an alpine valley to be served in his restaurant (Kofler 1978). Today, this formerly abundant species is rare on the floors of several alpine valleys (Landmann *et al.* 1999; Kyek and Maletzky 2006; Glaser 2008). The regulation of rivers along with settlements, agricultural fields and the construction of roads on limited space have rendered populated alpine valleys largely unsuitable for amphibians (Lippuner and Heusser 2001; Kyek and Maletzky 2006; Glaser 2008). At higher elevations in the Alps, habitat loss is less severe. Alpine amphibians nevertheless are threatened by the introduction into ponds and lakes of fish that feed on the eggs and larvae of amphibians and alter the composition of plankton assemblages (Schabetsberger *et al.* 1995, 2006).

Alluvial floodplains, the prime natural habitat for numerous amphibian species, have mostly disappeared in Austria (Tockner *et al.* 2006). Many amphibians need dynamic riverbeds and regular flooding for new breeding sites to form; this no longer occurs because of large-scale regulation of rivers along almost all of Austria's main watercourses (Pintar 2001; Wahringer-Löschenkohl *et al.* 2001); for an exception see Landmann and Böhm (2001).

As an additional threat, *Batrachochytrium dendrobatidis* (*Bd*), the causative agent of the amphibian disease chytridomycosis, was discovered in Austria in 2009 (Sztatecsny and Glaser 2011). A more extensive survey in 2010 showed that 50% of all 75 of the amphibian populations sampled throughout the country proved positive for *Bd* (Sztatecsny and Hödl 2011). None of the tested amphibians, however, showed signs of disease, making predictions of disease-associated effects difficult and warranting further investigations.

III. Austrian species of special concern for conservation

Many of Austria's amphibian species occur in various habitats from the lowlands to middle or high elevations in the Alps. Such generalist species may have suffered the greatest overall losses of habitat in terms of area, despite being able to survive in regions less affected by human activity. Distributionally restricted species, however, need to be given priority in terms of conservation. Austria's rarest amphibians are *L. helveticus*, which was found for the first time in 2008 (and therefore not included in Austria's Red List published in 2007) at a site close to the Swiss border (Grabher and Niederer 2011) and *B. calamita*, known from two populations close to the borders with Germany and the Czech Republic (Cabela *et al.* 2001). Both species, however, reach their distributional limits to the west and northwest of Austria (Gasc *et al.* 1997), and have rather large distributional ranges extending into other countries.

Triturus dobrogicus, *B. bombina* and *P. fuscus* are restricted to Austria's eastern lowlands, and the largest populations occur in alluvial forests (Cabela *et al.* 2001). The former two species are listed in Annex II of the EU Natural Habitats Directive, and important habitats are protected as parts of the Donauauen and Neusiedler See National Parks and as Natura 2000 conservation areas along the river Morava. Despite this local protection, these species suffer from riverine regulation that prevents regular flooding and the creation of new breeding sites (Pintar 2001). Accordingly, this has caused local populations to decline in otherwise preserved areas (Sztatecsny 2009). *Triturus carnifex* and *T. cristatus* also occur in Austria and both are listed in Annex II of the EU Natural Habitats Directive; they have a larger range than does *T. dobrogicus*, and all three species hybridize naturally (Jehle *et al.* 2011, Wielstra *et al.* 2013).

Bombina variegata also is listed in Annex IV of the EU Natural Habitats Directive. However, due to its habitat requirements (it breeds in small temporary bodies of water), conservation areas based on the directive may not overlap with its actual distribution (Kyek and Maletzky 2006). This species appears to be declining in particular on the northern edge of the Alps, where natural breeding sites have become rare (Gollmann and Gollmann 2002; Gollmann 2007). In the east of Austria, *B. variegata* overlaps with *B. bombina* and the two species form a hybrid zone (Szymura and Barton 1991).

Almost 20% of all *B. variegata* (n=124) and 34% (n=102) of *B. bombina* populations tested in Austria proved positive for *Bd*, with the maximum incidence being 57% for one *B. variegata* population (Sztatecsny and Hödl 2011) and 43% for metamorphs in one *B. bombina* population (Sztatecsny and Glaser 2011). The effects of *Bd* infection on the species of *Bombina* need to be investigated, since it may cause reduced survival of metamorphs (Garner *et al.* 2009).

Bufo viridis depends on open ruderal habitats that are becoming increasingly rare. Of special conservation concern are isolated populations of *B. viridis* in the Alps, where the species is restricted to the floors of warm, inner alpine valleys (Cabela *et al.* 2001; Gollmann 2007). Its habitat therefore overlaps with areas where there is intense human activity. This species is assumed to be extirpated in the province of Salzburg, and declining in Carinthia and Tirol (Cabela *et al.* 2001; Kyek and Maletzky 2006; Gollmann 2007)

Almost half the distributional range of *S. atra*, which is endemic to the Alps, lies within Austria (Gasc *et al.* 1997). There are few data on the abundance of *S. atra* in the Austrian Alps (see below for a new surveying approach), but there is evidence for declines at its distributional limit at lower elevations where human activity is intense (Cabela *et al.* 2001). At higher elevations, the alpine salamander still is assumed to be abundant (Gollmann 2007).

IV. Conservation measures and monitoring programmes

There is no national institution responsible for the protection of amphibians in Austria, and most activities are locally restricted. In each province, small monitoring programs are underway, with countrywide distributional data being gathered at the Natural History Museum in Vienna (http://www.nhm-wien.ac.at). The website of Austria's Herpetological Society (ÖGH, http://www.herpetozoa.at) provides a collection of links to local initiatives and monitoring programs. The Lower Austrian League for Nature Conservation (http://naturschutzbund.at) established a database of roadside drift fences to reduce road kills in the province of Lower Austria, as has already been accomplished in other provinces such as Carinthia (http://www.amphibienschutz.at) and Salzburg (http://www.herpag-hdn.amphibien.at).

In 2009 a countrywide survey was initiated with the aim of increasing knowledge about the distribution of *S. atra* and *S. salamandra* by involving the public (Reinthaler-Lottermoser *et al.* 2010). Citizen scientists can enter records of either salamander species in a Google map interface on the project's web blog (http://alpensalamander.eu/blog/); these are stored in a MySQL database after inspection by the research team. Although involvement of the public may lead to some misidentifications and incorrect records, it allows for quick and easy input and availability of data. In one year (2009–2010), 5,681 reports on alpine salamanders were gathered through the project's web blog (Reinthaler-Lottermoser *et al.* 2010), compared to 1,002 records collected between the years 1768 and 2010 and available on the database of Vienna's Natural History Museum (S. Schweiger, personal communication). Obviously, the application of a community-based approach is restricted to a small percentage of any local amphibian fauna, as only certain species are reliably identifiable by laypersons.

V. Conclusions

Loss of habitat is the greatest threat to Austria's amphibians. In the eastern lowlands, important amphibian habitats are protected by National Parks but regulation of rivers causes ongoing deterioration of these habitats. The floors of alpine valleys receive little protection and are among the most anthropogenically altered habitats in Austria, leaving little space for amphibians. At high elevations, the introduction of fish into naturally fish-free ponds and lakes is considered the greatest threat to amphibian populations. Diseases such as the panzootic chytridiomycosis, discovered in Austria in 2009, could further threaten already declining populations. To better organize conservation efforts in Austria, a coordination centre for the protection of amphibians is highly desirable.

VI. Acknowledgements

I thank W. Hödl and R. Jehle for their valuable comments on an earlier version of the manuscript. During writing of the manuscript, I was supported by the Austrian Science Fund (FWF) grant P22069.

VII. References

Auer, I., Böhm, R., Jurkovic, A., Lipa, W., Orlik, A., Potzmann, R., Schöner, W., Ungersböck, M., Matulla, C., Briffa, K., Jones, P., Efthymiadis, D., Brunetti, M., Nanni, T., Maugeri, M., Mercalli, L., Mestre, O., Moisselin, J.M., Begert, M., Müller-Westermeier, G., Kveton, V., Bochnicek, O., Stastny, P., Lapin, M., Szalai, S., Szentimrey, T., Cegnar, T., Dolinar, M., Gajic-Capka, M., Zaninovic, K., Majstorovic, Z. and Nieplova, E., 2007. HISTALP – historical instrumental climatological surface time series of the Greater Alpine Region. *International Journal of Climatology* **27**: 17–46.

Brehm, A.E. and Schmidtlein, R., 1902. *Brehms Tierleben*. Volume 3 in Kriechtiere, Lurche, Fische, Insekten, Niedere Tiere. Bibliographisches Institut, Leipzig.

Cabela, A., Grillitsch, H. and Tiedemann, F., 2001. *Atlas zur Verbreitung und Ökologie der Amphibien und Reptilien in Österreich*. Umweltbundesamt, Wien.

Frost, D.R., 2011 Amphibian Species of the World: an Online Reference. Version 5.5 (31 January, 2011) Electronic Database accessible at http://research.amnh.org/vz/herpetology/amphibia/: American Museum of Natural History New York, USA.

Garner, T.W.J., Walker, S., Bosch, J., Leech, S., Rowcliffe, J.M., Cunningham, A.A. and Fisher, M.C., 2009. Life history tradeoffs influence mortality associated with the amphibian pathogen Batrachochytrium dendrobatidis. *Oikos* **118**: 783–91.

Gasc, J., Cabela, A., Crnobrnja-Isailovi, J., Dolmen, D., Grossenbacher, K., Haffner, P., Lescure, J., Martens, H., Martinez Rica, J., Maurin, H., Oliveira, M., Sofianidou, T., Veith, M. and Zuiderwijk, A., 1997. *Atlas of Amphibians and Reptiles in Europe*. Societas Europaea Herpetologica and Museum National d'histoire Naturelle, Paris.

Glaser, F., 2008. Amphibien in inneralpinen Tallagen. *Bioskop* **04/08**: 33–9.

Gollmann, G., 2007. Rote Liste der in Österreich gefährdeten Lurche (Amphibia) und Kriechtiere (Reptilia). Pp. 37–60. in vol. 14/2 *Rote Listen gefährdeter Tiere Österreichs: Kriechtiere, Lurche, Fische, Nachtfalter, Weichtiere*, ed. by K.P. Zulka. Bundesministerium für Landwirtschaft und Forstwirtschaft, Umwelt und Wasserwirtschaft. Böhlau Verlag, Vienna.

Gollmann, B. and Gollmann, G., 2002. *Die Gelbbauchunke: von der Suhle zur Radspur*. Laurenti Verlag, Bielefeld.

Grabher, M. and Niederer, W., 2011. Der Fadenmolch *Lissotriton helveticus* (Razoumowsky, 1789), eine neue Amphibienart für Österreich. P. 7 in *UMG Berichte*. UMG Umweltbüro Grabher, Begrenz.

IUCN 2011. IUCN Red List of Threatened Species. Version 2011.2: www.iucnredlist.org. IUCN Red List of Threatened Species.

Jehle, R., Thiesmeier, B. and Foster, J., 2011. *The Crested Newt – A Dwindling Pond Dweller*. Laurenti-Verlag, Bielefeld.

Kofler, A., 1978. Zum Vorkommen von Reptilien und Amphibien in Osttirol (Österreich). *Carinthia II* **168**: 403–23.

Kyek, M. and Maletzky, A., 2006. *Atlas und Rote Liste der Amphibien und Reptilien Salzburgs. Stand Dezember 2005*. Naturschutz-Beiträge, Amt der Salzburger Landesregierung, Naturschutzabteilung, Salzburg.

Landmann, A. and Böhm, C., 2001. Amphibien in Gebirgsauen: Artenbestand, Laichplatzangebot und Laichplatznutzung durch Grasfrosch (*Rana temporaria*) und Erdkröte (*Bufo bufo*) in den Auen des Tiroler Lech. *Zeitschrift für Feldherpetologie* **8**: 57–70.

Landmann, A., Böhm, C. and Fischler, D., 1999. Bestandessituation und Gefährdung des Grasfrosches (*Rana temporaria*) in Talböden der Ostalpen: Beziehungen zwischen der Größe von Laichpopulationen und dem Landschaftscharakter. *Zeitschrift für Ökologie und Naturschutz* **8**: 71–9.

Lippuner, M. and Heusser, H., 2001. Geschichte der Flusslandschaft und der Amphibien im Alpenrheintal. *Zeitschrift für Feldherpetologie* **8**: 81–96.

Pintar, M., 2001. Die Amphibien der österreichischen Donauauen. *Zeitschrift für Feldherpetologie* **8**: 147–56.

Reinthaler-Lottermoser, U., Meikl, M., Gimeno, A., Weinke, E. and Schwarzenbacher, R., 2010. A new approach for surveying the alpine salamander (*Salamandra atra*) in Austria. *Acta Herpetologica* **5**: 249–54.

Schabetsberger, R., Jersabek, C.D. and Brozek, S., 1995. The impact of alpine newts (*Triturus alpestris*) and minnows (*Phoxinus phoxinus*) on the microcrustacean communities of two high altitude karst lakes. *Alytes (Paris)* **12**: 183–9.

Schabetsberger, R., Grill, S., Hauser, G. and Wukits, P., 2006. Zooplankton successions in neighboring lakes with contrasting impacts of amphibian and fish predators. *International Review of Hydrobiology* **91**: 197–221.

Stuart, S.N., Chanson, J.S., Cox, N.A., Young, B.E., Rodrigues, A.S.L., Fischman, D.L. and Waller, R.W., 2004. Status and trends of amphibian declines and extinctions worldwide. *Science* **306**: 1783–6.

Sztatecsny, M., 2009. *Die Amphibien der March-Thaya-Auen unter besonderer Berücksichtigung der Langen Luss: Bestand, Gefährdungsursachen und Maßnahmenkatalog.* Universität Wien, Vienna.

Sztatecsny, M. and Glaser, F., 2011. From the eastern lowlands to the western mountains: First recordings of the chytrid fungus *Batrachochytrium dendrobatidis* in wild amphibian populations from Austria. *The Herpetological Journal* **21**: 87–90.

Sztatecsny, M. and Hödl, W. 2011. Chytridiomykose in Österreich: Bestandsaufnahme einer tödlichen Amphibienkrankheit / Chytridiomycosis in Austria: survey of a deadly amphibian pathogen. Endbericht / Final report. Bund-Bundesländerkooperation.

Szymura, J.M. and Barton, N.H., 1991. The genetic structure of the hybrid zone between the fire-bellied toads *Bombina bombina* and *B. variegata*: Comparisons between transects and between loci. *Evolution* **45**: 237–61.

Tockner, K., Bunn, S.E., Quinn, G., Naimann, R.J., Stanford, J.A. and Gordon, C., 2006. Floodplains: critically threatened ecosystems. Pp. 45–61 in *The State of the World's Waters*, ed

N.C. Polunin. Cambridge University Press, Cambridge.

Wahringer-Löschenkohl, A., Baumgartner, C. and Pintar, M., 2001. Laichplatzverteilung von Amphibien in den niederösterreichischen Donauauen in Abhängigkeit von der Gewässerdynamik. *Zeitschrift für Feldherpetologie* **8**: 179–88.

Wielstra, B., Baird, A.B. and Arntzen, J.W. 2013: A multimarker phylogeography of crested newts (*Triturus cristatus* superspecies) reveals cryptic species. *Molecular Phylogenetics and Evolution* **67**: 167–75.

59 Conservation and decline of European amphibians: The Czech Republic

Lenka Jeřábková, Martin Šandera and Vojtech Baláž

I. Introduction
   A. Species list and recent changes
   B. Species' distribution and richness

II. Amphibian declines and species of special conservation concern
   A. Species of the Red List
   B. Conservation status reports
   C. Main causes of population declines

III. Conservation measures and monitoring programmes
   A. Monitoring and mapping of species

IV. Conclusions

V. References

Abbreviations and acronyms used in the text or references

EEA	*European Environment Agency*
IUCN	*International Union for Conservation of Nature*
KFME	*a system of squares for mapping the abundance of the Central European biota*
NCA CR	*Nature Conservation Agency of the Czech Republic*
NGO	*Non-governmental Organization*
SOD	*Species Occurence Database of NCA CR*
ZO ČSOP	*Czech Union for Nature Conservation*

I. Introduction

The Czech Republic is a medium-sized country (78,866 km^2) in Central Europe. Its topography is dominated by areas of small elevational diversity with *c.* 60% of the area below 500 m. Low mountains are situated along the border areas with the highest peak reaching 1,602 m. For centuries, the landscape has been intensively changed by agriculture, forestry, mining, urbanization and other human activities. At present, about 38% of the country's area is used as arable land, 34% is forested and almost 13% covered by permanent grasslands (http://eagri.cz). The temperate climate in the country is described as transitional between warm maritime and warm continental (EEA Report 2002).

Water from the country flows into three different seas: the North Sea (Elbe River), Baltic Sea (Oder River) and Black Sea (Morava River). Larger permanent, natural bodies of water are rare (a few glacial lakes, bog pools and oxbow lakes). The majority of free-standing bodies of water are man-made. Many lakes emerged as a result of open-cast mines and quarries being filled with water after operations ceased. Ponds for aquaculture have been an important part of the Czech landscape since the 12th century. In many rural areas, a small pond represented an integral part of each village's plan.

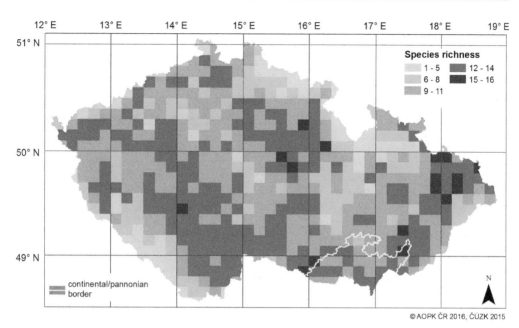

Fig. 59.1   Geographic variation of the species richness of species within the Czech Republic. The unit is the number of species per square of KFME mapping grid 10′ lat × 10′ long. Data from map at a scale of 1.1.2000.

The area of the Czech Republic encompasses two biogeographical regions: the majority of the country falls within the Continental region, and the southeastern corner (Southern Moravia) is part of the Pannonian region (EEA Report 2002) (Figure 59.1).

A. Species list and recent changes

The total number of amphibian species present in the Czech Republic is 21. No changes in known biodiversity of amphibians have been made since the 1990s, when three caudate species (*Lissotriton helveticus*, *Triturus dobrogicus* and *Triturus carnifex*) were found in the country.

Some of the species (*Epidalea calamita*, *Lissotriton montandoni*, *L. helveticus*, *T. dobrogicus* and *T. carnifex*) are present in populations on the margin of their distributional range. Therefore, their Czech populations have increased susceptibility to any negative processes (Dufresnes and Perrin 2015). The majority of Czech amphibian species are listed in the Red List for the Czech Republic as well as Council Directive 92/43/EEC on the Conservation of Natural Habitats and of Wild Fauna and Flora (Table 59.1). No knowledge of a viable population of non-native amphibian species is known from the country.

B. Species' distributions and richness

The terrain influences the local biodiversity of amphibians despite the small elevational and climatic variability of the country. Lowlands and plains are the most species-rich areas (Figure 59.1). The likely limiting factor in areas over 800 m is, however, the reduced variability of bodies of water, not climatic conditions. The presence of preferred habitats and climatic conditions are the main factors influencing the local occurrence of broadly distributed species such as *Bufo bufo*, *Rana temporaria*, *Hyla arborea* or *Lissotriton vulgaris*. Almost one third of the 21 species living in the Czech Republic is present only on the margin of their distributional range.

Lissotriton helveticus was discovered and documented in the Czech Republic in 1990 (Janoušek and Smutný 1990; Zavadil and Kolman 1990). This species is represented only by a few easternmost local populations that occupy pools close to the western border between the Czech Republic and Germany.

Table 59.1   Conservation or legally protected status of amphibian species in the Czech Republic according to individual legislative actions evaluations of conservation.

CR – Critically endangered; **EN** – Endangered; **VU** – Vulnerable; **NT** – Near threatened as indicated by the Red List of the Czech Republic (Jeřábková *et al.* 2016), Directive 92/43/**EEC** (listed in Annex **II, IV, V**), Act No 114/1992 Coll. or: **CE** – Critically endangered; **SE** – Severely endangered; **E** – Endangered. Nomenclature according to Frost (2016).

		Red List of the Czech Republic	Directive 92/43/EEC	Act No 114/1992 Coll.
Fire-bellied toad	*Bombina bombina*	EN	II, IV	SE
Yellow-bellied toad	*Bombina variegata*	CR	II, IV	SE
Natterjack toad	*Epidalea calamita*	CR	IV	CE
Green toad	*Bufotes viridis*	EN	IV	SE
Common toad	*Bufo bufo*	VU		E
European tree frog	*Hyla arborea*	NT	IV	SE
Common spadefoot	*Pelobates fuscus*	NT	IV	SE
Common frog	*Rana temporaria*	VU	V	
Moor frog	*Rana arvalis*	EN	IV	CE
Agile frog	*Rana dalmatina*	NT	IV	SE
Edible frog	*Pelophylax* kl. *esculentus*	NT	V	SE
Pool frog	*Pelophylax lessonae*	VU	IV	SE
Eurasian marsh frog	*Pelophylax ridibundus*	NT	V	CE
Italian crested newt	*Triturus carnifex*	EN	II, IV	CE
Northern crested newt	*Triturus cristatus*	EN	II, IV	SE
Danube crested newt	*Triturus dobrogicus*	CR	II	
Carpathian newt	*Lissotriton montandoni*	CR	II, IV	CE
Palmate newt	*Lissotriton helveticus*	CR		CE
Alpine newt	*Ichthyosaura alpestris*	VU		SE
Smooth newt	*Lissotriton vulgaris*	VU		SE
Common fire salamander	*Salamandra salamandra*	VU		SE

At the opposite end of the country, stretching across the eastern part of Southern Moravia, lies the western distributional limit of *T. dobrogicus*, a species otherwise broadly distributed along the Danube river. Several isolated sites, inhabited in proximity to the confluent rivers Thaya and Morava, have been known only since 1993, when the presence of *T. dobrogicus* in the country was first discovered (Zavadil *et al.* 1994; Zavadil 1995). Several dozens of sites occupied by *T. carnifex* are located west of the area occupied by *T. dobrogicus* near the Thaya river. The latter species, with source populations in the Italian peninsula, was found in 1997 and was the last amphibian confirmed for the country (Piálek *et al.* 1998, 2000; Zavadil 1998; Piálek and Zavadil 1999; Horák and Piálek 2001). A species subendemic to the Carpathian Mountain range, *L. montandoni*, has its northwestern distributional limit in the Beskydy and Jeseníky mountains on the northeastern border of the country.

II. Amphibian declines and species of special conservation concern

The majority of amphibian species in the Czech Republic are becoming rare and several species may face local extirpation from the country in the near future. The species at highest risk are *T. dobrogicus*, *T. carnifex* and *L. helveticus*. Their vulnerability lies in the fact that they have always occupied very limited areas. A similar case is the isolated population of *L. montandoni* present at Jeseníky Mountain, which is isolated from the main population in the Carpathian mountain range. All these species, however, are broadly distributed and are generally less endangered outside the Czech Republic.

Epidalea calamita, Rana arvalis, Bombina bombina, the western population of *Bombina variegata,* and apparently *Triturus cristatus* have all experienced dramatic population declines in the past 50 years and have become rare, occurring in fragmented populations. If the trend continues, these species will soon be in danger of extirpation from the country. *Lissotriton vulgaris, Ichthyosaura alpestris,* the "Carpathian" populations of *L. montandoni, B. bufo, Bufotes viridis, R. temporaria* and *Pelophylax lessonae* are species that were common in the past, but are showing strikingly rapid declines in the past 20 years. *Pelobates fuscus* and *Salamandra salamandra* have always been scarce and localized but these species are not yet in danger of extirpation from the country (NCA CR 2016).

A. Species of the Red List
Several lists of conservation concern for amphibian species in the area of the Czech Republic have been continuously updated, based on the changing state of knowledge and formal criteria for categorization. The first Red List was made in 1981 (Baruš 1981) and from the 18 species then known, 12 were considered to be endangered. Subsequent lists were provided by Baruš *et al.* (1988), Baruš and Zima (1989), and Baruš (1989). The Red List from 2003 implemented the criteria of IUCN (Zavadil and Moravec 2003) and included all amphibian species present. The last update of the list was made in 2015 (Jeřábková *et al.* 2016, in press) and again evaluated all 21 species. None of the species is listed as extinct or extinct in the wild; however, the categories Critically Endangered, Endangered, and Vulnerable each contain five species while Near Threatened includes six species (see Figure 59.2). It is worth noting that these categorisations are not based on actual rareness of the species, but upon observed population trends.

The conservation status of amphibians in the Czech Republic, reflected through the scope of previous (Zavadil and Moravec 2003) and updated Red Lists (Jeřábková *et al.* 2016), has in general worsened. Over the period between these two publications, six species were rated as having become more endangered than previously (Table 59.2). *Epidalea calamita* is the species that is declining most rapidly. In many areas the density of this species has plummeted below the limit of detection and its whole population currently consists of a few isolated fragments. It is presently considered Critically Endangered (NCA CR 2016). *Bufo. viridis* was re-classified from Vulnerable to

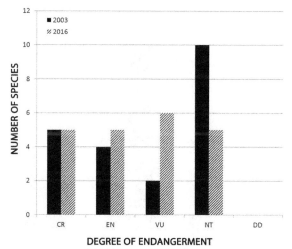

Fig. 59.2   Degree of endangerment of Czech amphibians in 2003 and 2016.

Categories of endangerment are those of the Red List of the Czech Republic (Jeřábková *et al.* 2016): **CR** = Critically endangered; **EN** = Endangered; **VU** = Vulnerable; **NT** = Near threatened; **DD** = Data Deficient

Table 59.2 Comparison of the conservation status of amphibians in the Czech Republic in 2003 and 2016. Details in Zavadil and Moravec (2003) and Jeřábková *et al.* (2016). **CR** = Critically endangered; **EN** = Endangered; **VU** = Vulnerable; **NT** = Near threatened

Species	RL 2003	RL2016	Species	RL 2003	RL2016
Bombina bombina	EN	EN	Pelophylax ridibunda	NT	NT
Bombina variegata	CR	CR	Rana temporaria	NT	VU
Bufo bufo	NT	VU	Salamandra salamandra	VU	VU
Epidalea calamita	EN	CR	Ichthyosaura alpestris	NT	VU
Bufotes viridis	NT	EN	Triturus carnifex	CR	EN
Hyla arborea	NT	NT	Triturus cristatus	EN	EN
Pelobates fuscus	NT	NT	Triturus dobrogicus	CR	CR
Rana arvalis	EN	EN	Lissotriton helveticus	CR	CR
Rana dalmatina	NT	NT	Lissotriton montandoni	CR	CR
Pelophylax kl. esculentus	NT	NT	Lissotriton vulgaris	NT	VU
Pelophylax lessonae	VU	VU			

Endangered. Population declines also have been observed in *B. bufo, R. temporaria, I. alpestris* and *L. vulgaris* and they are now classified as Vulnerable. Only *T. carnifex* was re-classified into a lower category, thanks to an increase in its population facilitated by intense management of habitats. It is now listed as Endangered, despite its very limited distribution (Jeřábková *et al.* 2016).

B. Conservation status reports

Reports on the main results of the surveillance under article 11 of Annex II, IV and V species are submitted to the European Commission in six-year cycles. The conservation status of monitored species is described by a scale of four levels: Favourable (FV), Inadequate (U1), Bad (U2), and Unknown). The Czech Republic has so far produced two reports, one in 2007 and the other in 2013 (Figure 59.3). In total, 16 species present in the Czech Republic are listed as Species of European Importance, and 13 of them live in both biogeographic regions within the country. The reports were based predominantly on comparisons between historical and current knowledge of the state of populations.

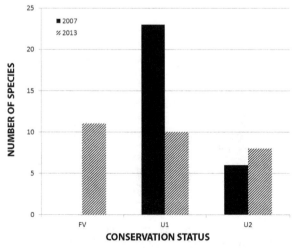

Fig. 59.3 Conservation status of Czech species of amphibians in 2007 and 2013.
Scale: **FV**=Favourable; **U1**= Inadequate; **U2**= Bad.

The two reports are an important source of information on the changes observed in the populations of amphibians in the country. The comparison shows an increase in species with a positive status of their populations. This can be explained, however, by an increased knowledge of species' distributions in the country, and not by actual changes of the populations. In conjunction with the Red Lists, the most endangered are some of the species near the edge of their distributions (Chobot 2016): *E. calamita*, *L. montandoni*, *T. carnifex* and *T. dobrogicus*. These species are threatened by loss of habitat. This is the main cause of decline even in the broadly distributed *T. cristatus*, *B. bombina*, *B. variegata*, *P. fuscus*, *R. temporaria* and *R. arvalis*. The species with observed growth of their populations or expansion of their distributions are *H. arborea*, *Rana dalmatina*, and the *Pelophylax* kl. *esculentus* complex. *H. arborea* seems to be resistant to many negative effects and in some areas is expanding its range. The increase and spread of *R. dalmatina* is likely triggered by the warming climate, as this species seems to be tolerant of rising temperatures and drought. It is continually expanding its distribution in the country and very dense populations inhabit several sites. The increase of *R. dalmatina* also may be enhanced by decreased competition with *R. arvalis* and *R. temporaria*, both of which are declining.

C. Main causes of population declines

The dynamic viewed as 'declines and expansions' of amphibian populations is a natural process triggered by random seasonal factors. However, it also has been influenced by human pressure in many forms. The long-term problems for the majority of amphibian species in the country are continuous fragmentation of their populations, as well as declines in population size and in the number of inhabited sites.

Degradation and loss of habitat is the most prominent cause of amphibian population declines and extirpations in Europe (Stuart *et al.* 2004). Serious threats to amphibians are the changes in water balance of the landscape caused by land-development of meadows and woods, canalization, straightening and deepening of watercourses, deposition of soil and waste in puddles, and transformation of marshes and wet meadows into production meadows or woodlands.

The pressure for increasing the productivity of land caused loss of extensively used semi-natural habitats, followed by declines of species inhabiting those habitats. The most severe changes in landscape took place after 1948 with the collectivization of agriculture and the socialization of privately owned land. This led to losses in diversity of the landscape, and caused degradation of many biologically diverse and valuable habitats. The use of arable land has further intensified since the political changes of the 1990s, with growth in the amount of applied herbicides, pesticides and mineral fertilizers, despite an actual reduction of the area of arable land (Czech statistical office, https://www.czso.cz).

Intensification of the use of carp ponds began in the 1980s and fish production per hectare has more than doubled since then (Czech fish-farmers association www.cz-ryby.cz). To increase productivity, the ponds are fertilized by manure; treated with lime; technically cleaned; and fish feed added on a grand scale. Overstocking and the stocking of non-native fishes, and the spreading of invasive fish species has caused many bodies of water to be insufficient to support amphibian populations. Fish-production ponds are drained and filled repeatedly over the year, often in spring, with no regard for on-going amphibian reproduction. The littoral zone with macro-vegetation in ponds is often lost through deepening, and by shading by trees grown on the banks, or by the activity of overstocked fish. Even bodies of water not used for aquaculture (natural swimming pools, flood dams, ponds in quarries, municipal fire-fighting reservoirs) are often spontaneously stocked by random fish.

Since the political and social changes in 1989, many areas transformed by mining were technically reclaimed, but this led to a complete loss of naturally developing communities on such sites

(Vojar *et al.* 2016). The contemporary Czech legislation on mining forbids the excavation of ground-water to the surface at mining sites; thus, the newly formed waterways and bodies of water are subsequently covered by soil. The technical reclamations of mining areas, quarries and spoil banks thus have become a problem for the conservation of amphibian species. Sky ponds, wetlands and marshes that spontaneously form after mining sites have been abandoned, are suitable, and by some species even preferred, sites for reproduction. However, these have often been lost through the process that was originally intended to help return the area into a natural state.

The role of alien and invasive species is most evident in the case of fishes, either by their pre-dation on larvae and adults of amphibians (topmouth gudeon *Pseudorasbora parva*; brown bullhead *Ameiurus nebulosus*; black bullhead *Ameiurus melas*) (Rozínek *et al.* 2013; Šandera 2015). The Amer-ican mink *Mustela vison*, racoon *Procyon lotor* and racoon dog *Nyctereoutes lotor* are among invasive mammals predating amphibians (Rozínek *et al.* 2013).

Bombina variegata is a species using ephemeral bodies of water, such as puddles in wheel tracks on unsupported roads or puddles forming in quarries. Reconstruction of forest roads and tracks, overgrowing of puddles by vegetation, and drainage of puddles are important factors in the decrease of this, and other, pioneering species. Depositing oil and waste in quarries or overgrow-ing of ephemeral bodies of water at such sites greatly affects *E. calamita*, *Rana* and *Pelophylax*, as well as *P. fuscus*. These species are most influenced by drainage of fish ponds during the frogs' reproductive period in spring; their egg masses are destroyed by the lowering of the water level (Chobot 2016). Populations of *R. temporaria* are sensitive to the disruption of their hibernation sites (small water-courses such as creeks, streams and channels) that takes place in areas of intense agriculture. Road-kills are gaining importance with increasing traffic; threatening mostly slow and non-agile animals, like caudates and toads.

Emerging diseases of amphibians are another serious threat. The amphibian chytridiomycosis is caused by two fungal pathogens, *Batrachochytrium dendrobatidis* (Longcore *et al.* 1999), which infects mostly anurans, and the recently described *Batrachochytrium salamandrivorans* that infects and kills caudates (Martel *et al.* 2013). *Batrachochytrium dendrobatidis* is present in most of Europe as well as in the Czech Republic (confirmed in 2008 by Civiš *et al.* 2012). The fungus so far has been detected in 10 species (Baláž *et al.* 2014; Baláž, unpublished data). Some individual cases of mor-tality were observed in *B. variegata* and *B. viridis* (Baláž *et al.* 2013). The infection, however, is most prevalent in the *P.* kl. *esculentus* complex. Despite this, there is no evidence or sufficient data supporting *B. dendrobatidis* as a factor of amphibian decline. The potential for *B. salamandrivorans* invading the area is viewed with great concern, but it has not yet been confirmed, either in captive or wild amphibians, in the Czech Republic (Baláž *et al* 2016). In the past, cases of local epidemics of myasis were reported and had a great impact on populations of *B. bufo* (Zavadil *et al.* 1997).

III. Conservation measures and monitoring programmes

All of the amphibian species are considered endangered at various levels of intensity and are therefore protected by several legislative norms and laws (Table 59.1). *Rana temporaria* and *T. dobrogicus* are not listed as specially protected, but general laws on the protection of nature and against cruelty to animals apply to them. Sixteen species are listed as of European importance and for six of them Special Areas of Conservation under the European Union's Habitats Directive were established. Other areas are protected as natural reserves, protected landscape areas, or national parks and wetlands of international importance. The law automatically protects some of the hab-itats used by amphibians as important landscape features (ponds, lakes).

Active conservation and management of habitat is carried out by various NGOs and by organ-isations contractually bound to the NCA CR or to administration authorities on all levels of governance. The most common conservation activities include building of permanent or temporary linear road barriers, construction and restoration of small bodies of water, management of sites at ponds by mowing or pasturage, removal of turf and eradication of undesirable species of fish. Alternative approaches to the management of habitats are used locally. Former military training areas that were found to have great biodiversity are managed by the military, using heavy off-road vehicles to produce desirable disturbance and renewal of fine-scale patterns of habitat.

There is no National Action and Management Plan in process for any critically endangered or endangered species, but one is being prepared at the time of writing of this chapter and is expected to begin sometime in 2019.

A. Monitoring and mapping of species

The first extensive surveys of amphibians have contributed to a species-distribution atlas (Moravec 1994). More intense studies began after the year 2000 owing to obligations in the formation of the NATURA 2000 network after the integration of the Czech Republic into the European Union. The initial surveys were aimed at the Czech species in Annex II of the Habitats Directive (*T. cristatus*, *T. carnifex*, *T. dobrogicus*, *L. montandoni*, *B. bombina* and *B. variegata*). Many new data on the distribution of other species were collected along the way. A fine-scale mapping scheme of amphibians (and reptiles) started in 2008 under the lead of NCA CR and in 2014 covered the whole extent of the country (and still continues). Additionally, a long-term monitoring of selected species in a representative network of localities has been carried out continuously every year from 2006 onward, with the aim of detecting long-term trends in populations and the factors contributing to them. Further information on the status of species is collected via faunal inventories, and appropriate assessments are made. A citizen-science scheme was set up whereby members of the public submit data by website BioLib (http://www.biolib.cz) or by mobile application BioLog (http://biolog.nature. cz). All the data that are collected are stored by NCA CR in a species-occurrence database (SOD) and are available to nature conservation authorities for better planning of their activities or, upon request, for scientists to use in their research.

IV. Conclusions

The country is inhabited by 21 amphibian species (13 anurans; 8 caudates) which is a relatively high diversity given the size of the country and the situation in surrounding countries (Sillero *et al.* 2014). However, the majority of the species are in decline. The main factors behind this decline are linked to loss of habitat, e.g. intense changes in landscape use, over-utilization of pesticides and herbicides, inappropriate use of bodies of water, lack of biologically wise management of habitats. The amphibian chytrid fungus, *B. dendrobatidis*, infects several species and is widely present, but with no apparent effects on populations.

Long-term surveillance of amphibians' distributions and population dynamics is predominantly carried out by employees and contractual workers from NCA CR; significant help also is received from volunteers. The factors directly linked to population changes are checked as a part of monitoring and mapping programmes. All data are collected in SOD and used by NCA CR and nature-conservation authorities.

The first country-wide program for the specific enforcement of conservation for a population is being prepared for a single species, *E. calamita*. Other species are under protection by Special Areas of Conservation, national parks, protected landscape areas and nature reserves. Despite some positive results in previous years and the legal protection of all species, more active conservation efforts are needed to halt the on-going decline and disappearance of amphibians and their habitats in the Czech Republic.

V. References

Baláž, V., Kubečková, M., Civiš, C., Rozínek, R. and Vojar, J., 2013. Fatal chytridiomycosis and infection loss observed on wild-infected toads in captivity. *Acta Vet Brno* **82**: 351–5.

Baláž, V., Vojar, J., Civiš, P., Šandera, M. and Rozínek, R. 2014. Chytridiomycosis risk among Central European amphibians based on surveillance data. *Diseases of Aquatic Organisms* **112**: 1–8.

Baláž, V., Solský, M., Jelínková, A., Havlíková, B., Rozínek, R. and Vojar, J., 2016. Výzkum trojice nejzávažnějších patogenů obojživelníků v České republice (The surveillance of three most dangerous pathogens of amphibians in Czech Republic). In J. Bryja J., F. Sedláček and R. Fuchs R. (eds.): *Zoologické dny České Budějovice, 2016.* Sborník abstraktů z konference 11.–12. února 2016 [in Czech].

Baruš, V., 1981. Návrh seznamu ohrožených taxonů obratlovců (Vertebrata) fauny ČSSR. *Vertebratologické zprávy* **1981**: 35–42 [in Czech].

Baruš, V., Donát, P., Trpák, P., Zavázal, V. and Zima, J., 1988. *Red Data list of vertebrates of Czechoslovakia. Přírodovědné Práce Ústavu Československé.* Akademie Věd Brno **22**: 1–33 [in Czech].

Baruš, V., Bauerová, Z., Kokeš, J., Král, B., Lusk, S., Pelikán, J., Sládek, J., Zejda, J. and Zima, J., 1989. Červená kniha ohrožených a vzácných druhů rostlin a živočichů ČSSR Vol. 2. Kruhoústí, ryby, obojživelníci, plazi, savci. Státní zemědělské nakladatelství, Praha [in Czech].

Baruš, V. and Zima, J. 1989. Červený seznam kruhoústých, ryb, obojživelníků, plazů a savců ČSSR. In: Záchranné chovy a odchovy, Sborník referátů, Nový Jičín 1987, pp. 300–5, Praha & Nový Jičín [in Czech].

Civiš, P., Vojar, J. and Baláž, V. 2012. Current state of *Bd* occurrence in the Czech Republic. *Herpetological Review* **43**: 150–9.

Chobot, K., [ed.], 2016. Druhy a přírodní stanoviště. Hodnotící zprávy o stavu v České republice 2013. Agentura ochrany přírody a krajiny ČR & Ministerstvo životního prostředí, Praha [in Czech].

EEA. 2002. European Environment Agency Report 1/2002. Europe's biodiversity, biogeographical regions and seas. www.eea.europa.eu.

Dufresnes, C. and Perrin, N. 2015. Effect of biogeographic history on population vulnerability in European amphibians. *Conservation Biology* **29**: 1235–41.

Frost, D.R. 2016. Amphibian Species of the World: an Online Reference. Version 6.0 (accessed September 2016). Electronic Database accessible at http://research.amnh.org/herpetology/amphibia/index.html. American Museum of Natural History, New York, USA.

Horák, A. and Piálek, J., 2001. Genetic Structure of the *Triturus cristatus* superspecies in the Czech Republic. In: P. Lymberakis, E. Vlakos, P. Pafilis and M. Mylonas, M. [eds.], *Herpetologia Candiana*, pp. 89–92, SEH & Natural History Museum of Crete, University of Crete, Irakleio.

Janoušek, K. and Smutný, Z., 1990. Čolek hranatý *Triturus helveticus* novou součástí herpetofauny Československa. Akváriumterárium, Praha, **39 (9)**: 30–32.

Jeřábková, L., Krása, A., Zavadil, V., Mikátová, B. and Rozínek, R., 2016. Červený seznam obojživelníků a plazů České republiky, pp. 83–93. In: Plesník, J., Hanzal, V., Brejšková, L. (eds.): *Červený seznam ohrožených druhů České republiky, Obratlovci.* Příroda, Praha, 22, in press [in Czech].

Longcore, J.E., Pessier, A.P. and Nichols, D.K. 1999. *Batrachochytrium dendrobatidis* gen. et sp. nov., a chytrid pathogenic to amphibians. *Mycologia*, **91**: 219–27.

Martel, A., Spitzen-van der Sluijs, A., Blooi, M., Bert, W., Ducatelle, R., Fisher, M.C., Woeltjes, A., Bosman, W., Chiers, K., Bossuyt, F. and Pasmans, F. 2013. *Batrachochytrium salamandrivorans* sp. nov. causes lethal chytridiomycosis in amphibians. *Proceedings of the National Academy of Sciences of the United States of America* **110**: 15325–9.

NCA CR (2016). *Species Occurence Database of NCA CR.* [on-line database; portal.nature.cz] [cit. 2016–01–31].

Piálek, J., Zavadil, V. and Reiter, A. 1998. Presence of the Italian Crested Newt *Triturus carnifex* in the Czech Republic I. Morphological evidence. In: 9th Ordinary General Meeting Societas Europaea Herpetologica, Chambéry, France, 25–9 August 1998.

Piálek, J. and Zavadil, V. 1999. A new newt species, *Triturus carnifex* for the Czech Republic. In Biennial report, 1997–8, pp. 32–4, Institute of Vertebrate Biology, Academy of Sciences of the Czech Republic, Brno.

Piálek, J., Zavadil, V. and Valíčková, R. 2000. Morphological evidence for the presence of *Triturus carnifex* in the Czech Republic. *Folia Zoologica* **49**: 33–40.

Rozínek, R., Fischer, D. and Baláž, V. 2013. Impact of invasive species on the herpetofauna of the Czech Republic. In: 17th European Congress of Herpetology Veszprém, Hungary, p. 49.

Sillero, N., Campos, J., Bonardi, A., Corti, C., Creemers, R., Crochet, P.A., Isailovic, J.C., Denoel, M., Ficetola, G.F., Goncalves, J., Kuzmin, S., Lymberakis, P., de Pous, P., Rodriguez, A., Sindaco, R., Speybroeck, J., Toxopeus, B., Vieites, D.R. and Vences, M. 2014. Updated distribution and biogeography of amphibians and reptiles of Europe. *Amphibia-Reptilia* **35**: 1–31.

Stuart, S.N., Chanson, J.S., Neil, C.A., Young, B.E., Rodrigues, A.S.L., Fischman, D.L. and Waller, R.W. 2004. Status and Trends of Amphibian Declines and Extinctions Worldwide. *Science* **306**: 1783–6.

Šandera M. 2015. Zvýšený výskyt skokanů skřehotavých s deformovanými končetinami u Staré Lysé. (Increased number of marsh frogs with deformed limbs at the village of Stará Lysá). *Vlastivědný zpravodaj Polabí* **44**: 4–19 [in Czech].

Vojar, J., Doležalová, J., Solský, M., Smolová, D., Kopecký, O., Kadlec, T. and Knapp, M. 2016. Spontaneous succession on spoil banks supports amphibian diversity and abundance. *Ecological Engineering* **90**: 278–84.

Zavadil, V. and Kolman, P. 1990. Čolek hranatý novým druhem naší fauny. *Živa* **38**: 224–7.

Zavadil, V., Piálek, J. and Klepsch, L. 1994. Extension of the known range of *Triturus dobrogicus*: electrophoretic and morphological evidence for presence in the Czech Republic. *Amphibia-Reptilia* **15**: 329–35.

Zavadil, V. 1995. Čolek dunajský *Triturus dobrogiucus* (Kiritzescu, 1903) novým druhem obratlovce České republiky. *Ochrana Přírody* **50**: 18–20 [in Czech].

Zavadil, V., Kolman, P. and Marik, J. 1997. Frogs myiasis in the Czech Republic with regard to its occurrence in the Cheb district and comments on the bionomics of *Lucilia bufonivora* (Diptera, Calliphoridae). *Folia Facultatis Scientiarum Naturalium Universitatis Masarykianae Brunensis Biologia* **95**: 201–210.

Zavadil, V. 1998. Problémy v ochraně obojživelníků a plazů v České republice. In: I. Otáhal and J. Plesník (eds.), *Záchranné programy živočichů v České republice.* Pp. 60–3, ZO ČSOP Nový Jičín. Stanice pro záchranu živočichů Bartošovice na Moravě [in Czech].

Zavadil, V. and Moravec, J. 2003. Červený seznam obojživelníků a plazů České republiky. Pp. 83–93. In: J. Plesník, V. Hanzal and L. Brejšková (eds.), *Červený seznam ohrožených druhů České republiky, Obratlovci.* Příroda, Praha, **22** [in Czech].

60 Amphibian declines and conservation in Slovakia

Ján Kautman and Peter Mikulíček

Abbreviations and acronyms used in the text and references

IUCN	*International Union for the Conservation of Nature*
m a.s.l.	*metres above sea level*

I. Introduction

Slovakia is a country in central Europe where climatic and environmental conditions are influenced by the presence of two important European geographical units: the Carpathian Mountains covering mainly central and northern Slovakia and the Pannonian (Carpathian) Lowland in the southern part of the country. The territory of Slovakia is characterized by a high range in elevation, 94–2,654 m a.s.l. in a small area (49,036 square kilometres), which results in rapid drainage of water and an absence of natural standing waters, with the exception of alpine lakes of glacial origin. Amphibians have been recorded up to 2,000 m a.s.l. (Zavadil 1993). The most important areas, with respect to amphibian species diversity and abundance, are situated in the Borská, Podunajská and Východoslovenská lowlands. Despite extensive agriculture and meliorations in the past, these lowland regions still retain natural wetlands associated with the Morava, Danube and Latorica rivers, which provide suitable habitats for amphibian reproduction and larval development. However, gradual drainage, accumulation of soil, and reduction of small natural bodies of water, especially in alluvial regions, result in an acute shortage of reproductive habitats, and species with lower mobility are gradually disappearing from the country. Over the past few decades, many changes have occurred in the country and populations of almost all amphibian species have been significantly reduced.

II. Amphibian species in Slovakia

Eighteen amphibian species are distributed in Slovakia (Lác 1963; Oliva *et al.* 1968; Baruš *et al.* 1992). No introduced species occur yet and no native species have yet been extirpated. Below, we summarize basic information about particular species, with emphasis on their protection.

A. Family Salamandridae, true salamanders and newts, mlokovité

1. Salamandra salamandra (Linnaeus, 1758), fire salamander; salamandra škvrnitá

The Slovak population of this species occurs within the area near the northern border. It occurs in deciduous forests of all of the Slovak mountains. With increasing elevation and in coniferous forests the abundance of species is significantly lower. This species is absent in the agricultural landscape and in the lowlands. Hypsometric distribution is at elevations 150–1,000 m a.s.l. The species is dependent on the preservation of the forest environment where there are refugia and food supplies. Breeding habitats represented by springs, small streams and wetlands free of fish are crucial for the survival of this species. Its existence may be affected by insensitive management of forests and water, trout breeding, and more recently significant warming, which causes desiccation of many reproduction sites. Owing to its bionomy, the fire salamander is considered to be a less vulnerable, stabilized species.

2. Triturus cristatus (Laurenti, 1768), great crested newt, mlok hrebenatý

This species has penetrated Slovakia from surrounding countries where it is much more numerous. It occurs in mountain ranges at elevations of 200–900 m a.s.l. Occasionally it spreads towards the southern border and when in contact with *Triturus dobrogicus* it creates hybrid populations. Breeding sites are various: still, deep (0.5–2 meters) bodies of water without fish. An acute shortage of suitable breeding habitats, caused by gradual degradation and destruction by agriculture and forestry, succession or fish culture, as well as by isolation of populations, has resulted in *T. cristatus* now being the rarest amphibian in Slovakia. Its populations are small and vulnerable.

3. Triturus dobrogicus (Kiritzescu, 1903), Danube crested newt, mlok dunajský

In Slovakia this species reaches the northwestern border of its range. It occurs mainly in the lowlands from an elevation of 100 to 250 m a.s.l. In uplands and foothills, it meets and hybridizes with *T. cristatus*. The absolute majority of sites are concentrated at an elevation of 100–200 m a.s.l. in the Borská, Podunajská, and Východoslovenská lowlands. Populations in the lowlands of central Slovakia are scattered. Breeding habitats are represented by deeper, still waters, without fish. This species lives in the woods and in a mosaic landscape, where it finds sufficient refugia for the terrestrial stage close to breeding sites. Many suitable sites, especially near the major rivers, were destroyed by water-management and agricultural activities, especially by melioration and by chemical pollution. The species has long been declining and it is one of the rarest and most endangered species.

4. Ichthyosaura alpestris (Laurenti, 1758), alpine newt, mlok horský

The Slovak populations of this species are located in the northeastern part of its distributional range. It occurs mainly at middle and higher elevations in the mountains of Central and Eastern Slovakia from 300 m to 1,850 m (predominantly between 500 and 1,200 m a.s.l.). It is a typical montane forest species with most of its sites in the highest mountains. Breeding habitats are represented by a diversity of still waters of various depths. It avoids waters in which fish occur. Survival of the species is locally influenced by insensitive silvicultural techniques and by the drainage of soil. Shallow waters recently have been reduced by lack of rainfall and by high temperatures. Under such changes the species' populations have declined in the long term.

5. Lissotriton montandoni (Boulenger, 1880), Montandon's newt, mlok karpatský

Slovak populations of this species are situated in the northwestern part of its range. The species occurs mainly at medium and high elevations in the mountains in northern Slovakia, mainly in coniferous and beech forests at elevations of 500–1,000 m a.s.l. It occasionally descends into the basins and rarely into open country. Breeding habitats are shallow wetlands with still water or the littoral areas of overgrown deeper waters. Melted snow fens or depressions in roads are often sufficient for reproduction also. This species avoids waters inhabited by fish. Survival is affected mostly by drainage and insensitive silvicultural techniques. With increasing climatic warming, more and more sites dry out during the year, and this significantly affects the breeding success of increasingly isolated small populations. At the rare sites where this species occurs syntopically with *L. vulgaris*, these two species hybridize.

6. Lissotriton vulgaris *(Linnaeus, 1758), smooth newt, mlok bodkovaný*

Slovak populations of this species are located in the middle of its distributional range. The Smooth newt occurs throughout the whole territory, especially in the lowlands; it also spreads through basins to the middle elevations. In mountainous areas it is rare, only occasionally reaching elevations up to 1,000 m. The frequency of occurrence decreases with increasing elevation. Suitable sites for reproduction are still waters with rich aquatic vegetation. This species prefers mosaic and open country and is missing in cold, densely forested mountainous areas. Survival of the species is influenced mostly by insensitive management of agriculture and water, and by chemical pollution, fish breeding, and drying caused by global warming. Its abundance is dependent on the frequency of suitable sites for reproduction; these slowly are being destroyed.

B. Family Bombinatoridae, fire-bellied toads, kunkovité

1. Bombina bombina *(Linnaeus, 1761), fire-bellied toad, kunka červenobruchá*

This species occurs in the lowlands and hills of southern Slovakia, rarely exceeding elevations of 350 m a.s.l. At elevations from 200 to 500 m a.s.l., where it meets *Bombina variegata*, the two species hybridize, forming a continuous hybrid zone from the western to the eastern borders of Slovakia, particularly in the uplands and the foothills of the higher mountains. Fire-bellied toads prefer lowland meadows and forests where they breed in still waters, often also in seasonal waters and rain pools. Drainage, chemical pollution and agriculture, as well as recent drying of sites, are causing a sharp decline in their abundance, especially in the western part of the country.

2. Bombina variegata *(Linnaeus, 1761), yellow-bellied toad, kunka žltobruchá*

This species occurs in the uplands and mountains at elevations of 250–1,200 m a.s.l. Up to 500 m a.s.l. it often hybridizes with *Bombina bombina*. It prefers smaller bodies of water, such as seasonal pools, wheel tracks in roads, flooded pits and quarries. In the southern, warmer areas it is bound to the forests; in the colder northern part of Slovakia it occurs also in basins, pastures and peatbogs. This species and its habitats are declining slightly. Intensive management of forests damages native habitats, while heavy machines secondarily produces new ones. The declining trend in abundance is affected, in particular by the regulation of small streams that prevents the formation of new alluvial sites, but also by intensive agriculture and its associated chemical pollution. In recent, warm years unsuccessful reproduction because of the drying of breeding sites has become a serious problem.

C. Family Bufonidae, true toads, ropuchovité

1. Bufo bufo *(Linnaeus, 1758), common toad, ropucha bradavičnatá*

Often numerous populations of this species are distributed from the lowest to the highest mountains up to 2,000 m a.s.l. The species is most abundant in the warmer forests of plains and lower mountains where there are plentiful suitable reproductive habitats. This toad is a migrating eurytopic species, preferring deciduous forests, rarely occurring also in deforested and agricultural landscapes, and often in municipalities. It breeds in ponds, dams, wetlands, seasonal pools, streams and garden pools. It is also able to reproduce in waters inhabited by fish. It is still abundant throughout the country, but its abundance varies or is reduced at some sites. As a migratory species it is threatened by mortality on roads and by the presence of barriers to migration.

2. Bufotes viridis *(Laurenti, 1768), green toad, ropucha zelená*

In Slovakia this species lives outside forested areas, especially in steppe habitats, but also inhabits mosaic, agricultural and disturbed landscape. It is often found in villages, particularly in the lowlands and hills. In the north, it spreads through deforested valleys up to an elevation of 800 m a.s.l. The green toad is an adaptable, xerophilous, pioneer species of the open country, villages, sand pits and quarries. For reproduction it frequently uses shallow, seasonal pools and puddles, often of anthropogenic origin. In densely populated areas, it is threatened by road traffic, the presence of barriers to migration, and extreme desiccation of the sites where it reproduces.

D. Family Hylidae, treefrogs, rosničkovité

1. Hyla arborea *(Linnaeus, 1758), common treefrog, rosnička zelená*

The Common treefrog occurs in lowlands and hills of southern Slovakia and it spreads via valleys to northern Slovakia where it is very rare. It also has been recorded at an elevation above 1,000 m a.s.l. For reproduction it requires small, still waters with rich vegetation and banks densely overgrown by shrubs and trees. It prefers sunny and warm habitats with high humidity. The species is gradually declining at many sites owing to draining of the country. Subsequent forestry and agricultural use, associated with increased chemical pollution, as well as regulation of waterways, has led to destruction of suitable breeding sites. On the other hand, the species rapidly inhabits newly created sites and spreads to areas from which it was previously unknown.

E. Family Pelobatidae, spadefoot toads, hrabavkovité

1. Pelobates fuscus *(Laurenti, 1768), common spadefoot, hrabavka škvrnitá*

This species occurs in the lowlands and hills of southern Slovakia, especially in open country and open, lowland forests. It has spread northward through basins along major rivers. It occurs rarely at elevations above 500 m a.s.l. It requires loose, sandy soils. Breeding sites are permanent, still waters surrounded by vegetation. Common spadefoots are considered rare because of their nocturnal, cryptic habits, but in places with suitable conditions mass occurrences have been observed. The species is threatened by land management, especially by chemical pollution, road traffic and loss of breeding habitat. The abundance of this species is gradually decreasing.

F. Family Ranidae, true frogs, skokanovité

1. Rana arvalis *(Nilsson, 1842), moor frog, skokan ostropysk ý*

The moor frog is the rarest anuran species in Slovakia; its range is restricted to the southwestern and southeastern parts of the country. A third small population penetrates from Poland through basins between the high mountains. This species occurs mainly in wet, waterlogged habitats and breeds in shallow, still waters, predominantly in natural undisturbed wetlands. Its abundance has been drastically reduced in the past few decades because it is very sensitive to environmental changes and chemical pollution. Due to significant damage of its native habitat, it is seriously threatened in Slovakia. Global warming, associated with drying of the breeding sites, is a major threat to many populations. Protection and conservation of this species should focus mainly on the preservation of natural habitats and reduction of negative anthropogenic influence.

2. Rana dalmatina *(Bonaparte, 1840), agile frog, skokan štíhly*

This species occurs both in the lowlands and uplands, especially in the warmer areas of deciduous forests and in steppe habitats in the southern part of Slovakia. In the central and northern parts of the country it extends through basins along large rivers. This is an adaptable, undemanding terrestrial species that breeds in still waters of various types and sizes. It is a species with stable populations, but locally threatened by the degradation of breeding sites. Climatic warming even helps this species to spread northward and to higher elevations. Given its adaptability, particularly to xerothermic conditions, it is not seriously threatened.

3. Rana temporaria *(Linnaeus, 1758), common grass frog, skokan hnedý*
The common grass frog occurs in all of the mountain ranges of Slovakia up to an elevation of 2,000 m a.s.l. It is most abundant in the mixed and coniferous forests of higher, cooler mountains. In warm deciduous forests it is rare and in the lowlands and deforested country it is absent altogether. It is a migratory, cryophilic woodland species. Various bodies of water constitute its breeding sites or, in the absence of these, it can breed in seasonal puddles or streams. The species still occurs in large numbers throughout its range, but it is significantly declining. The biggest threat is the loss of breeding sites, road traffic during migration and the construction of anthropogenic barriers (e.g. highways). Its conservation should focus mainly on preserving natural habitats and reducing negative anthropogenic impacts.

4. Pelophylax ridibundus *(Pallas, 1771), marsh frog, skokan rapotavý*
In Slovakia, this species is bound to the lowlands (up to an elevation of 300 m a.s.l.) and inhabits still waters and slow streams. Unlike *Pelophylax lessonae*, it prefers larger bodies of water, well saturated with oxygen and without rich submerged vegetation, often near big rivers. In suitable habitats it can live syntopically with the other two species of water frogs. Compared with *P. lessonae* it is often found in man-made habitats, such as gravel pits and channels. Human activities leading to the creation of artificial habitats supports the dissemination of this species at the expense of *P. lessonae*, which prefers more natural habitats.

5. Pelophylax lessonae *(Camerano, 1882), pool frog, skokan krátkonohý*
The pool frog is found in the lowlands and uplands to an elevation of 700 m a.s.l. Unlike *P. ridibundus* it prefers smaller, cooler and more densely vegetated still waters. It is the most terrestrial of all Central-European species of water frogs. It inhabits swamps, wetlands, mires, oxbow lakes, edges of ponds and reservoirs. In southern Slovakia it occurs mostly in forest wetlands and depressions, and alluvia of the larger rivers, which tend to be cooler than other bodies of water. It regularly occurs syntopically with the hybrid taxon *Pelophylax esculentus*; occurrence conjointly with *P. ridibundus* is rare. It is threatened by destruction of natural wetlands and peat-bogs, particularly in the agricultural landscape. It is disappearing from disturbed habitats where it is being replaced by *P. ridibundus*.

6. Pelophylax *kl.* esculentus *(Linnaeus, 1758), edible frog, skokan zelený*
This taxon originates by hybridization between the parental species *P. ridibundus* and *P. lessonae*. In Slovakia it occurs mainly in the lowlands and at middle elevations up to 700 m a.s.l. Most frequently it occurs syntopically with *P. lessonae*, less often with both parental species. It is relatively common. It inhabits various types of habitats like marshes, oxbows, sand pits, lakes and canals. It is rare along the banks of rivers and in large bodies of water, such as gravel pits. In addition to diploid individuals, triploid males have been found in western Slovakia. The genetic structure of hybrid populations of *P.* kl. *esculentus* in Slovakia is poorly known.

III. Species of special conservation concern
The newts *T. cristatus, T. dobrogicus, L. montandoni*, and the frogs *P. lessonae* and *R. arvalis* are rare and have a restricted or patchy distribution in Slovakia. Although they can locally reach high densities, generally they are less abundant than other amphibian species. The distribution and abundance of *Pelophylax lessonae* is negatively influenced by alteration of natural habitats to those made by humans (e.g. canals, gravel-pits, dams), where this species is replaced by its ecologically more plastic congener *P. ridibundus* (Mikulíček *et al.* 2015). Another taxon that merits special

conservation concern is the hybrid *P. kl. esculentus*, which is abundant in suitable habitats throughout Slovakia, but whose triploid form is geographically restricted, having been recorded only from several localities in western Slovakia. These triploid hybrids are males possessing two genomes of *P. lessonae* and one genome of *P. ridibundus* (LLR); they form diploid sperm with a *P. lessonae* genome (Mikulíček *et al.* 2015; Pruvost *et al.* 2015). Genotypic composition and a mode of gametogenesis make LLR triploids from western Slovakia unique among different types of *P. kl. esculentus* hybrids in Europe.

IV. Monitoring programmes

Monitoring of amphibian species of European importance in Slovakia (*B. bombina, B. variegata, B. viridis, H. arborea, P. fuscus, R. arvalis, R. dalmatina, R. temporaria, P. kl. esculentus, P. lessonae, P. ridibundus, T. cristatus, T. dobrogicus, L. montandoni*) follows implementation of the Habitats Directive of the European Union. It was conducted in the years 2013–2015 (Janák *et al.* 2015). Despite the short duration of this monitoring, it became obvious that the status of most monitored species is inadequate or bad. This is true also for the abundance of studied populations as well as assessments of the quality of habitats they occupy. The details of distribution and monitoring of amphibians in Slovakia are presented on web page www.biomonitoring.sk.

V. Red List of amphibian species in Slovakia

The Red List of amphibian species of Slovakia was summarized by Kautman *et al.* (2001) following the IUCN Red List Categories valid in the year 2001 (version 3.1). According to the Red List (Table 60.1) amphibians in Slovakia are assigned to the categories Endangered (EN, three species), Vulnerable (VU, five species), and Lower Risk (LR, ten species) with subcategories conservation dependent (LR: cd, five species), near threatened (LR: nt, three species) and least concern (LR: lc, two species).

Table 60.1   Red list of amphibian species of Slovakia according to the IUCN categories. The IUCN National Red List evaluates the status of species in the whole country; IUCN Carpathian Red List evaluates the status of species only in the Slovak Carpathians.

Species	IUCN National Red List	IUCN Carpathian Red List
Bombina bombina	LR: cd	VU
Bombina variegata	LR: cd	NT
Bufo bufo	LR: cd	NT
Bufotes viridis	LR: cd	NT
Hyla arborea	LR: nt	NT
Pelobates fuscus	LR: cd	VU
Rana arvalis	VU	EN
Rana dalmatina	LR: lc	LC
Rana temporaria	LR: lc	NT
Pelophylax kl. *esculentus*	LR: nt	NT
Pelophylax lessonae	VU	EN
Pelophylax ridibundus	EN	VU
Salamandra salamandra	LR: nt	LC
Ichthyosaura alpestris	VU	VU
Triturus cristatus	EN	CR
Triturus dobrogicus	EN	VU
Lissotriton montandoni	VU	VU
Lissotriton vulgaris	VU	NT

The conservation status of amphibian species occurring in the Carpathian region of Slovakia (IUCN Carpathian Red List) was summarized by Urban and Kautman (2014). One species is classified as critically endangered (CR), two species as endangered (EN), six species as vulnerable (VU), seven species as near threatened (NT), and two species are assigned to the category least concern (LC).

VI. Conclusions

Changes in diversity of amphibians result mainly from direct and indirect anthropogenic interventions and changes. Many species in Slovakia do not inhabit contiguous territory and their occurrence is becoming increasingly fragmented. Individual populations are isolated and cease to communicate with each other. The most negative effects on the distribution of amphibians are draining of the soil, use of insecticides and other chemicals in agriculture and forestry, regulation of flows of water, building of barriers to migration, fisheries management and, in recent years, significant climatic warming and its associated dry years. These factors have a significant influence on the quantitative decline of many species, most of which now have to be included among the seriously vulnerable. Constantly decreasing number of suitable breeding sites and, last but not least, a significant loss of insects and invertebrates as a basic food supply also contribute to a significant decline in amphibian populations. The incidence and mapping of chytridiomycosis, a global infectious disease, in Slovakia is still in early stages. Its agent, the chytrid fungus *Batrachochytrium dendrobatidis* was recently confirmed in Slovakia in the genera *Bombina, Bufotes, Pelophylax, Triturus* and *Lissotriton* (R. Rozínek, personal communication). Considering all of these threats, the prospects for the future are not optimistic for almost any species of amphibian.

VII. Acknowledgements

We are grateful to D. Jablonski (Comenius University in Bratislava) and P. Puchala (State Nature Conservancy of the Slovak Republic) for their helpful comments concerning distribution and legislative protection of amphibians in Slovakia.

VIII. References

Baruš, V., Král, B., Oliva, O., Opatrný, E., Rehák, I., Roček, Z., Roth, P., Špinar, Z. and Vojtková, L., 1992. *Obojživelníci – Amphibia*. Academia, Praha.

Janák, M., Černecký, J. and Saxa, A. (eds), 2015. *Monitoring živočíchov európskeho významu v Slovenskej republike*. Výsledky a hodnotenie za roky 2013–15. Štátna ochrana prírody Slovenskej republiky, Banská Bystrica.

Kautman, J., Bartík, I. and Urban, P., 2001. Červený (ekosozologický) zoznam obojživelníkov (Amphibia) Slovenska. In: Baláž, D., Marhold, K., Urban, P. (eds), Červený zoznam rastlín a živočíchov Slovenska. *Ochrana prírody* **20 (Supplement)**: 146–7.

Lác, J., 1963. Obojživelníky Slovenska. *Biologické práce* **9**: 1–76.

Mikulíček, P., Kautman, M., Kautman, J. and Pruvost, N.B.M., 2015. Mode of hybridogenesis and habitat preferences influence population composition of water frogs (*Pelophylax esculentus* complex, Anura: Ranidae) in a region of sympatric occurrence (western Slovakia). *Journal of Zoological Systematics and Evolutionary Research* **53**: 124–32.

Oliva, O., Hrabě, S. and Lác, J., 1968. *Stavovce Slovenska I. Ryby, obojživelníky a plazy*. Vydavateľstvo SAV, Bratislava.

Pruvost, N.B.M., Mikulíček, P., Choleva, L. and Reyer, H.-U., 2015. Contrasting reproductive strategies of triploid hybrid males in vertebrate mating system. *Journal of Evolutionary Biology* **28**: 189–204.

Urban, P. and Kautman, J., 2014. Carpathian Red List of Threatened Amphibians (Lissamphibia). In: *Carpathian Red List of Forest Habitats and Species, Carpathian List of invasive Alien Species* (draft). ŠOP SR Banská Bystrica, BioREGIO Carpathian project, 209–13.

Zavadil, V., 1993. Vertikale Verbreitung der Amphibien in der Tschechoslowakei. *Salamandra* **28**: 202–22.

61 Conservation status of amphibians in Norway

Leif Yngve Gjerde

Abbreviations and acronyms used in the text and list of references

DN	*Directorate for Nature Management*
EEA	*European Economic Area*
EFTA	*European Free Trade Association*
EIA	*Environmental Impact Assessments*
EU	*European Union*
NFHF	*Fieldherpetological Forum 2016*
NGO	*Non-governmental Organization*
NINA	*Norwegian Institute for Nature Research*
NIVA	*Norwegian Institute for Water Research*
NØBI	*Nordre Øyeren Biological Station*
NTNU	*Norwegian University of Science and Technology*

I. Introduction

A. Geography, landscape and climate

Norway covers an area of 323,895 square kilometres and stretches 1,752 kilometres from Lindesnes (57°59′N) in the south to Kinnarodden (71°08′N) in the north (Gjessing and Ouren 1983; Anonymous 1990, 1991). Within this area there are low strandflats, extensive fjords, high mountains (highest 2,469 m), hill lands, valleys and lowlands (Wallen 1970). Five of Köppen's climatic types occur within this area. A small area along the southern coast belongs to the Nemoral Zonobiome. Larger parts of southern Norway, east of the northeast-southwest orientated Caledonian mountain chain, belong to the Boreal Zonobiome and Boreonemoral Zonoecotone. An Interzonal Orobiome covers a larger part of southern Norway, while the Tundra Zonobiome is present in northern Norway. The western coast of Norway has high precipitation with mild winters and cool summers (Oceanic and Suboceanic sections). The Svalbard Islands have a costal tundra climate and glaciers cover large areas. The archipelago covers an area of 62,700 square kilometres.

B. Previous research

The first review of the Norwegian amphibian fauna was probably carried out in 1878 by Robert Collett, when he lectured at the Science Academy in Oslo, and was later published by him (Collett, 1879). Later reviews include Collett (1918), Johnsen (1935), Ruud (1949), Støp-Bowitz (1950), and Johnsson and Semb-Johansson (1992). Surveys of spawning sites were first carried out by Enger (1970) in the Fredrikstad area. Later surveys include the areas of Kongsrudmyra (Gjerde 1984, 1992), the Øyeren delta (Gjerde 1989a, 1991b, 1996), Kløfta (Rygg 1989), Romerike (Fossen *et al.* 1989) and Stange municipality (Aaseth *et al.* 1993).

Professor Arne Semb-Johansson studied *Bufo bufo* on three islands on the Hvaler archipelago every summer for 27 years beginning in 1966. In addition to monitoring the populations he also made notes on their morphology, and suggested their reduced body size constituted an insular adaptation (Semb-Johansson 1989). Dag Dolmen at the NTNU University Science Museum in Trondheim carried out his work on newts in central Norway from 1971 until he retired in 2014 (e.g. Dolmen 1982). Leif Gjerde from the Nordre Øyeren Biological Station initiated a survey and monitoring project on brown frogs at the Øyeren delta in 1988, which is still in progress (Gjerde 1989, 1996, in prep. b). During more recent years a number of theses on the occurrence of *Triturus cristatus* at Geitaknottane has been produced by students at the University of Bergen (Gjerde, in prep. a).

In Norway there have traditionally been only a few people working with amphibians. A *Norwegian Fieldherpetological Forum* was established as a network in 1991, and formalized into an organization in 2016 (see web page: www.nfhf.info/norway). Further information about amphibians in Norway is also available from www.herptiler.no.

C. Current status of amphibians

Norway has a relatively poor diversity of amphibian species. Norway and Sweden combined constitute the Scandinavian Peninsula, which has a land bridge to Finland/Russia only north of the Arctic Circle. There were connections to continental Europe (Denmark/Germany) during postglacial periods (7,500–6,000 B.C.) when the Baltic Sea was a lake (Lake Ancylus), but the land bridge disappeared before all species were able to disperse to the north. Since then, Norway has had a number of warm and cold climatic periods, and it is believed that the small Norwegian population of *Rana lessonae* is a relict from such a warm period. Seven amphibian species occur in Norway: two newts *Triturus cristatus* (Laurenti, 1768) and *Triturus vulgaris* (L., 1758); one toad *Bufo bufo* (L., 1758); and four frogs *Rana temporaria* L., 1758, *Rana arvalis* Nilsson, 1842 (Figure 61.1), *Rana lessonae* Camerano, 1882, and *Rana esculenta* L., 1758 (not native). No amphibian species are known from the Svalbard Islands, Bear Island or Jan Mayen.

Fog (1995) suggested that three colour varieties of *Rana* arvalis from Denmark (*striata, maculata, nigromaculata*) are in part geographic variations linked to habitat and with interspecific competition with *R. temporaria*. Furthermore, he described morphological variations linked to their distribution. For Eastern Europe he mentioned *unicolor* and *punctata* as additional examples of colour morphs. However, the ventral side was not described. In Norway only the colour morph *maculata* has been recorded. However, Gjerde (1993) mentioned a characteristic white stripe on the throat (but not on the belly), which is formed by the lack of blue colouration (100% of the males) (Figure 61.2 left) or marbled surface (94% of the females) (Figure 61.2 right), described as *oeyereniensis*. This feature

Fig. 61.1   A male *Rana arvalis* forma *oeyereniensis* from the Øyeren delta in 2011. Note its blue lek colouration. Photograph by Leif Yngve Gjerde.

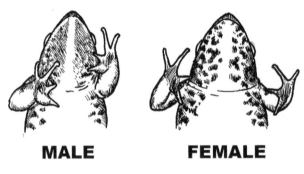

MALE   **FEMALE**

Fig. 61.2   *Rana arvalis* forma *oeyereniensis* from the Øyeren delta showing the characteristic stripe on the throat in the majority of individuals. The stripe is formed by lack of the blue colouration in males and by lack of marbling in females. Illustrations by Petter Bøckman.

also has been described by Elmberg (1978) from Umeå and Gotland (both in Sweden) (Nilson and Andrén 1981). However, the sample from Umeå included only one individual, and on the individuals from Gotland the entire belly was marbled as well. Forma *oeyereniensis* has been reported from Denmark, West Germany, Poland and mainland Sweden (13–35%), but the colouration of the belly has not been described for specimens from these regions. The status of the variety found in the Øyeren delta has yet to be assessed.

II. Status and distribution

A. Surveys

Dolmen (1982) published a temporary atlas based on work he initiated in 1973. It was mainly about newts in central Norway, but was expanded to include all amphibians throughout Norway. A number of other preliminary publications have appeared since. Beginning at the end of the 1980s interest in amphibians has increased, and a number of local and regional surveys have been carried out (e.g. Strand 1994; Gjerde 1997a, b, 2002; Spikkeland 1998; Hansen 1999).

B. Mapping biological diversity

In the Parliamentary instructions of 1997 (Anonymous 1997) it was required that all municipalities should carry out mapping and evaluate the biological diversity within each municipality boundary by 2003. The Regional Environmental Agencies provided 50% of the finances, and the municipalities were expected to provide the rest. However, implementation varied from place to place; some municipalities initiated their own surveys, but most only compiled existing knowledge into a report. These reports also included data, if available, on the status of amphibians within the specific geographic scope of the municipality. Some municipalities only provided databases.

III. Threats

None of the Norwegian amphibian species, except for *R. lessonae* (Anonymous 2006), are threatened by extirpation on a national scale. Most landscape features remain intact and are ecologically sustainable. This includes coastal, montane and forested areas. On a regional basis, amphibian populations are threatened in the agricultural landscape by loss of habitat, industrialized agriculture and destruction of ponds.

Within the boreal forests of eastern and central Norway, forestry management has a large impact on amphibian habitats by drainage of peat bogs. The introduction of fish to small bodies of water is also a general but extensive threat.

NØBI produced in 1994 some simple guidelines for management of amphibian populations (Gjerde 1994a).

A. Habitat and occurrence

Due to the earlier glaciations of Norway, a number of lakes and peat bogs, of all sizes, exist (Figure 61.3). Both in mountainous and boreal regions, there are a number of natural habitats still intact. In addition to these, the wetlands are connected to rivers and their flood plains.

Fig. 61.3 Annual spring flood at Årnestangen (the Øyeren delta). A typical spawning habitat (*Carex* swamp) for *Rana arvalis*. Photograph by Leif Yngve Gjerde.

In rural areas, amphibians are traditionally found in artificial ponds on farm estates. Usually a farm had one or two ponds functioning as a source of water for use by people and animals, and in case of fire. A new law came into effect in 1958 (Schei and Zimmer 1996: LOV-1957-05-31-1) which required all farmers to fence their ponds for security reasons. This, in combination with a change from production of milk and meat to the raising of cereals in the agricultural industry, resulted in the filling in of most ponds.

B. Red List

The concept of the Red List was developed and placed in use by the IUCN as early as 1964. Subsequently, it was adapted by a number of NGOs throughout the world. In Norway the first Red List was published in 1974 by the World Wildlife Fund (Hagen *et al.* 1974). The first governmental Red List was produced in 1978 by the Nature Conservation Inspector for South Norway and *Statens Naturvernråd* (the Governmental Nature Conservation Council) published a Red List in 1984 (Anonymous 1984). In 1988 the *Directorate for Nature Management* (DN) published a Red List (Christensen and Eldøy 1988). The Red List of 1992 attracted some controversy since it was published both by Sandlund and Størkersen, resulting in two separate lists from the same authorities. In 1999 the Red List was updated by Størkersen.

In 2005 the *Norwegian Biodiversity Information Center* was established by the *Directorate for Nature Management*. Today it is a part of the *Ministry of Education and Research*. It revised the status of the Norwegian amphibian species twice (Kålås *et al.* 2006, 2010) (Table 61.1). It is important to be aware that Red Lists made by management authorities often have political objectives influencing the choice of categories. This becomes especially apparent when NGOs and independent scientists have not been asked for advice. In Sweden this problem is solved by using scientists who have no connection with the management authorities (Ahlén and Tjernberg 1996).

C. Agricultural landscapes

Typical agricultural landscapes of Norway, where crops are grown, lie below the late-glacial marine limit. They are old sea beds with deposited clay. These areas are located at Jæren (southwestern Norway), Ringerike and Vestfold (southwest of Oslo), Romerike/Østfold (southeast to northeast of Oslo), and around Norway's largest lake, Lake Mjøsa (in the Hamar and Gjøvik regions), as well as in areas around Trondhjem Fjord. The landscapes in these areas have been altered by intensive industrial agriculture with extensive levelling of ground in ravine valleys.

Before 1940, agriculture was mainly conducted by manual labour, and included a variety of animals. It was common to have a pond as a source for drinking water and in case of fire. Furthermore, a number of ponds in grazing areas supplied horses, cattle, sheep and goats with drinking water.

Table 61.1   Status of threatened Norwegian amphibians. **CR** = Critically endangered; **NT** = Near threatened; **VU** = Vulnerable

Species	Kålås *et al.* (2010)	Status NFHF [1]	Threats
Rana arvalis	NT	VU	Drainage of peat bogs; ploughing of floodplains and wetlands; road constructions isolating populations.
Rana lessonae	CR	CR	Only one metapopulation of 100–500 adults.
Triturus cristatus	VU	VU	Destruction of ponds on farmland; drainage of peat bogs; ploughing of floodplains and wetlands; road constructions isolating populations.
Triturus vulgaris	NT	VU	Destruction of ponds on farmland; drainage of peat bogs; ploughing of floodplains and wetlands; road constructions isolating populations.

[1] Norwegian Fieldherpetological Forum, 2016.

In 1957 a law was introduced to prevent children from drowning. It required that all ponds be reported to the police, and be secured by fencing (Schei and Zimmer 1996: LOV-1957-05-31-1). Since farming in Norway was gradually changing from animal production to the growing of crops, and with the access of water through pipes, ponds became less important. Accordingly, an increasing number of ponds were filled in during the following decades and the space they had occupied used for other purposes. This law was cancelled in 1995, and integrated into the Norwegian Planning and Building Act. Now a pool was defined as any open water with a depth of over 20 centimetres, and the municipality (not the police) became responsible for inspection.

Today most areas have few remaining ponds, e.g. in Fet municipality only two remained by the year 2000 (Gjerde 2002), while a school project in Ullensaker municipality located 122 ponds on recent maps, only 83 of which still persist (Rygg 1989). Spydeberg municipality had over 200 ponds on 300 farms in the 1930s. By 1984 there remained only 30–50% of them, and ten years later most of those were gone.

D. Forested regions

Most of the mountain chain in southern Norway consists of the Paleic Peneplain. As most of this area is relatively flat, the density of variably sized bodies of water is very high. Also peat bogs are quite frequent. Most regions east of the Caledonian mountain chain also have a high density of pools and lakes of various sizes. Except for residential and agricultural regions, this area is mostly covered with boreal forest of mountain birch (*Betula pubescens tortuosa*), birch (*Betula pubescens pubescens*), pine (*Pinus sylvestris*) and spruce (*Picea abies abies*).

Recreational fishing is a popular hobby in Norway, and the *Norwegian Association of Hunters and Anglers* is a strong organization promoting angling. Fish (mostly trout, *Salmo trutta*) are artificially introduced and cultivated in most lakes (e.g. Ingierd 2006). This includes small bodies of water down to 80 x 80 m. Newts are sensitive to predation by fish and, consequently, few localities remain suitable for these salamanders.

E. Introduced species

In 2003, Petter Bøckmann told the author that he had received rumours of a new frog species on the western coast of Norway. It transpired that it was a botanist, Svein Imsland, who first discovered the frogs. Subsequently, Mikaelsen (2008a, b) published the information that R. esculenta and R. lessonae had been introduced to the island Finnøy, just outside Stavanger. A Norwegian family had brought 20–30 individuals of mixed species from their vacation in Poland during the spring of 2003. Since then, the frogs have spread to a number of ponds on the island. In 2008 the species had established themselves at 4 of 7 lakes and 5 of 18 ponds (Dolmen, 2009). Potential threats of introduced species of amphibians to native ones are predation, competition and transfer of disease.

IV. Current law and protection

A. Protection of species

The law on hunting and capture of wild game came into effect in 1982 (Anonymous 2014b: LOV-1981-05-29-38). Until then, plant or animal species needed to be protected specifically if they were under threat. With this new law it widened the definition of "game" to include all species of amphibians, reptiles, birds and mammals, prohibiting general hunting, unless a designated hunting season was specified for the species. This reversed the hunting philosophy in favour of the species – the "mirror prinsipp". Similar laws came into effect in a number of European countries at this time. In essence, these gave a wide protection to all amphibian species, but not to their habitats.

B. Protection of biological diversity

In 2009 an additional law was introduced, the biodiversity act (Naturmangfoldloven). The aim of this law was to protect biological diversity and ecological processes by sustainable management, and for the benefit of human activity, culture, health and wellbeing, both at present and in the future. The objective for the conservation of each species is the long-term protection of its genetic variation, and that the species exist in sustainable populations within their natural geographic range. To obtain this objective the entire environment needs to be protected to allow all the species' ecological functions to proceed. This law came into effect on 1 July 2009 (Anonymous 2014c: LOV-2009-06-19-100).

C. Environmental impact assessments (EIA)

Regulations on environmental impact assessments have existed since 1990, but have been revised a number of times since. New regulations pursuant to the *Norwegian Planning and Building Act* came into effect on 1 January 2015. The objectives of these regulations are to ensure that consideration for the environment and society is included during preparation of measures/changes, and to decide if, and on what terms, those measures/changes are implemented.

It is usually the municipality's case worker who assesses the need for any environmental impact assessment (screening). If found necessary, usually a general biological study is carried out, usually based on existing knowledge. In most cases, however, current knowledge is too poor, fragmented and accidental to give any correct picture of the specific amphibian fauna at a given area. This leads to wrong and misguided conclusions. A field study is usually required.

The implementation of environmental laws integrated into construction work like road building and construction of wind turbines is relatively new, and no tradition has been developed for good planning and mitigation in the interest of environmental conservation.

The forest industry has developed guidelines and ethical standards that accompany the law, but nothing includes amphibians. The road authorities, however, have developed standards for environmental impact assessments and mitigations through guidelines like "Roads and Wildlife" (Iuell 2005). The implementation and integration of wildlife laws into the management of projects has become increasingly common (although the laws require full compliance from when the laws came into effect). This is a process that takes time. Most projects today consider environmental laws during the planning phase, but a field evaluation of the effects of any mitigation is still rare.

D. Mitigation

Although regulations on environmental impact assessments have existed for 28 years, implementation by authorities of sectors (e.g. municipalities, road authorities) has gradually become adopted over time. Even today, however, the importance of EIA is still trivialized. The change does not depend on regulations alone, but follows the knowledge, practice and respect every project coordinator has for the value of nature. What is essential is that mitigation proposals need to be included in the EIA. This includes mitigation both during construction and post-construction. Even if the law provides clear recommendations to post-construction EIA in which the effect of mitigation is tested, and even improved, such measures will never happen unless included in the EIA and in the budget for the construction work.

EIA has become relatively common in most sectors, but mitigation for amphibian populations is still relatively new. Only four mitigations have been carried out on amphibian passages under roads, and all during the past 12 years (Strand and Stornes 2007; Strand *et al.* 2009). Unfortunately, these have not been completely successful owing to lack of experience and neglect of effective designs from other countries, so all require post-construction improvements.

Table 61.2   Reviews of amphibian tunnel projects in Norway.

Locality	Mitigation	Reference	Faults
Volda	1997/2010	Strand *et al.* (2009)	Experimenting with untested designs for tunnel
Nesodden	2006	Strand (2001b); Strand and Stornes (2007)	Experimenting with untested designs for tunnel
Merkja, Nordre Øyeren	2015	Strand (in preparation)	Experimenting with untested designs for tunnel. The effect of tunnels not evaluated.

An overview of all Norwegian mitigation projects for amphibians is almost impossible owing to the numerous sector authorities involved. No central database or library on the topic exists. We know replacement ponds have been used for road projects and in connection with farming (e.g. Heggland 2010). However, it is the amphibian tunnels that receive the eye of the media, making it possible to get a proper overview of most projects (see Table 61.2). The latest and most extensive mitigation for amphibian populations was carried out in Merkja (Figure 61.4; Nordre Øyeren Nature Reserve) during 2014. This project is used as an example of obvious mistakes that have been made (Gjerde 2009, in prep. c):

- Recommendations ignored: noise screening, stone deposits along road for hibernation, creating additional ponds
- No dialogue between EIA scientists and mitigation
- Lack of replacement habitats (both terrestrial and aquatic)
- Concrete not coated, resulting in excessively high pH

Fig. 61.4   Amphibian tunnel at Merkja (Nordre Øyeren Nature Reserve) illustrating extensive foraging areas out of reach for amphibians. The prevention of communication between researchers from the EIA and mitigation stages of the road resulted in unnecessary mistakes. Also, there was a lack of a buffer zone between the terrestrial and aquatic habitats. The fence and tunnel should have been placed higher up on the slope. Mitigation failed to include replacement habitats. Photograph by Leif Yngve Gjerde.

- Entrances to tunnels on one side below ground level
- Poor drainage of tunnels
- No post-construction study of the success of the tunnel

The biggest mistake was that during EIA the scientists hired for construction work and post-construction testing were different individuals with no exchange of information. This resulted in important information being lost during the process. The second mistake was that details and information concerning the mitigation should have been included in the EIA. Failure to do so resulted in information being lost and ignored later in the building process.

V. Monitoring programmes and conservation measures

The Parliament has given instructions on monitoring biological diversity in Norway (Anonymous 1997). The Directorate for Nature Management wish to include amphibians in a national monitoring programme and they have charged the Norwegian Institute for Nature Research (NINA) with assessing the possibility for such a scheme. NINA published their recommendations in 2000 (Hårsaker *et al.* 2000). Such a program has still to be initiated, however, and the ongoing non-governmental bodies have not been consulted.

A. Population Monitoring

The first monitoring of amphibians in Norway was carried out by Professor Arne Semb-Johansson on three islands of Hvaler (outer Oslo Fjord) over a 24-year period (Semb-Johansen 1989). It started in 1966 and included a 2.2-km walking transect repeated 10–20 times each summer (Semb-Johansson 1992).

A monitoring project on brown frogs (Figure 61.5) was initiated by Nordre Øyeren Biological Station in 1988 (see Figure 61.6). The objective was to monitor population sizes of *R. temporaria*

Fig. 61.5   Individuals of *Rana arvalis* from the Årnestangen metapopulation in the Øyeren delta during April 2011 showing their blue lek colouration. The spawning period lasts only for a few days and the blue colouration takes 1–2 days to develop after initiation of calling at the spawning site, during which the blue colour may be only partially developed or lacking (Gjerde 1993). Photograph by Leif Yngve Gjerde.

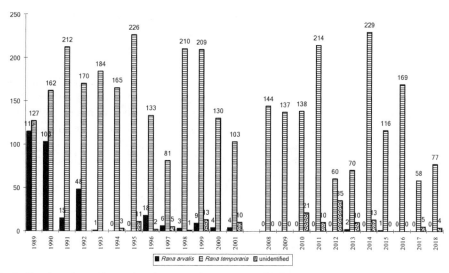

Fig. 61.6 Number of reproducing females at pond No. 20 in Nordre Øyeren Nature Reserve, monitored by the author. The work began in 1989, and is still being carried out by Nordre Øyeren Biological Station (NØBI). The project was discontinued from 2002–2007. Today we know that data from single ponds give limited information owing to short-term fluctuations in the population. For this reason, the project was gradually extended to include a number of spawning sites within the same metapopulation. Monitoring changes in habitat and general status of the breeding population (sink, medium, source) in relation to distances between suitable surrounding habitats (spawning and foraging sites) have been shown to be much more relevant in the author's experience.

and *R. arvalis* by registering the number of egg clusters in the Øyeren delta just 20 km east of Oslo. Since each female lays only one cluster of eggs, the numbers represent the population size of reproducing females (Figure 61.6). This is an accurate and effective way of monitoring populations. The 1988 season started as a trial, but the number of localities has increased to allow for monitoring three metapopulations, not individual ponds (Gjerde 1994b, in prep. b).

B. Phenology

In connection with the monitoring of metapopulations of brown frogs in Nordre Øyeren, data on timing of spring activity (phenology) were also collected, especially the date of the first cluster of eggs. Minimum and maximum ambient temperatures are compared with rainfall to assess the beginning of each spawning season (Gjerde 1996c, in prep. b). These data are important when assessing short-term climatic changes.

C. Repeated surveys

Except at Nordre Øyeren, there is no real monitoring going on in Norway. A program has been suggested by the authorities, but lack of funding and cooperation so far has prevented its implementation. However, a number of distributional surveys have been carried out by measuring the occurrence/absence of species at spawning localities. Such studies have been repeated to measure changes of occurrence and distribution through time (e.g. Gjerde 2008a, b; Strand 2001a).

D. Protected areas

The Danish conservation law protects all natural bodies of water of 100 m² or larger. In Norway, however, there is no general law specifically protecting amphibian habitats. Norway is not a member of the EU, but has cooperation and obligations as an EFTA (European Free Trade Association) country, through the EEA agreement. This implies that most legislation within the EU, is usually integrated into Norwegian law. However, the EU Habitats Directive has never been legally integrated into Norwegian law, and it has not been implemented in Norway. From 2001 onwards, however, a new water-resource law came into effect (Anonymous 2014a), protecting all natural waterways. The definition of a waterway includes semi-terrestrial areas covered by flood water at some time during a 10-year period (maximum 10-year average flood level). This covers areas/ habitats potentially important for *R. arvalis*. However, implementation of the law has proved ineffective and has been ignored by local authorities such as Fet municipality in the Nordre Øyeren Nature Reserve (Figures 61.7, 61.8). Thus, the only real protection of areas important to amphibian populations is carried out by the creation of Nature Reserves where entities other than amphibians are important aspects of the protection process.

So far, two areas in which amphibian populations have been a crucial argument for the legal process have been protected. They are Kongsrudmyra and Geitaknottane.

Kongsrudmyra (Figure 61.9) is located just 30 km east of Oslo and consists of a valley 2 km long and surrounded by agriculture to the north and east, and boreal forest to the south and west. The entire length of the valley's floor is covered with peat bogs, with a small lake in the centre. Buffer areas surrounding the bog include swamp forests of birch (*Betula pubescens pubescens*) and spruce (*Picea abies*). Three small lakes surrounded by small bogs are located in the boreal forest to the west of the ridge above the valley. The first studies on amphibians were carried out during spring and summer of 1983 (Gjerde 1984). A follow up study was carried out in 1990 and 1992 (Gjerde 1992) and revealed the occurrence of all five Norwegian amphibian species. Furthermore, the population of *T. cristatus*, *R. arvalis* and *R. temporaria* were relatively high. This led to a number of organizations getting involved to protect the area. The area was temporarily protected on

Fig. 61.7   The picture is from construction work during 2014 in Merkja (Nordre Øyeren Nature Reserve), illustrating the loss of wetland. The road authorities ignored the water-resource law of 2001, and mitigation did not include the replacement of lost wetlands. Photograph by Leif Yngve Gjerde.

Fig. 61.8 A pond within the Nordre Øyeren Nature Reserve (Ramsar site). It has the largest population of *Rana temporaria* within the reserve, but has been threatened a number of times by road construction, digging of ditches, run-off from agriculture and the ploughing of wetlands. The picture shows a normal spring flood in an area that since 2001 has been protected by the water-resource law. The picture is from 2006 and illustrates the landowners ploughing below the defined wetland boundary. Poor management and violations of the law resulted in the extirpation of *Rana arvalis* in the 1990s, although in some subsequent years a few individuals have reappeared and reproduced. Photograph by Leif Yngve Gjerde.

Fig. 61.9 Kongsrudmyra (viewed from north to south). This area is known for its many habitats within a small area, thus resulting in a high species diversity of plants, insects, amphibians and birds. Photograph by Leif Yngve Gjerde.

16 February 1996, and permanently protected on 13 December 2002 as a forest reserve with its special qualities of valuable amphibians.

Geitaknottane is located on a peninsula between the Hardanger and Bjørna Fjords, 30 km southeast of Bergen. The landscape is rocky and includes areas below and above the tree line. The area was first mentioned by Dolmen (1993). Torstein Solhøy and Anders Hobæk at the University of Bergen have since supervised over eight theses and students' reports from this area (Myklebust 1998; Hage 1999; Hobæk *et al.* 2000; Gutiérrez 2002; Berge 2005; Strømme 2005; Paulsen 2006; Paulsen *et al.* 2009), making it the most extensively studied area in Norway. The metapopulation of *T. cristatus* is large and included 1,680 individuals in 180 ponds surveyed in 2008 (Paulsen *et al.* 2009). An estimate for the total area suggests a population of 3,000–10,000 individuals (Myklebust 1998; Hage 1999; Gutiérrez 2002; Paulsen 2006). These populations are considered to be some of the largest in the world (Steinvåg 2010) and are certainly the largest in Norway. An area 13.7 km^2 in size at Geitaknottane was protected as a Nature Reserve on 19 December 1997.

VI. References

Aaseth, H., Bekken, J. and Ødegaard, R., 1993. Undersøkelser av salamanderforekomst i dammer og tjern i Stange 1992. Report no. 3/93. Fylkesmannen i Hedmark, Miljøvernavdelingen.

Ahlén, I. and Tjernberg, M., 1996. *Rödlistade ryggradsdjur i Sverige.* Sveriges Lantbruksuniversitet, Uppsala.

Anonymous, 1984. *Truete planter og dyr i Norge.* Statens naturvernråd, Tønsberg.

Anonymous, 1990. Index to names on 1:50 000 scale maps of Norway Series 711, Volume I. Forsvarets karttjeneste.

Anonymous, 1991. Index to names on 1:50 000 scale maps of Norway Series 711, Volume III. Forsvarets karttjeneste.

Anonymous, 1997. Miljøvernpolitikk for en bærekraftig utvikling. Stortingsmelding 508 (1996–7).

Anonymous, 2006. Handlingsplan for damfrosk *Rana lessonae.* Rapport 2006–2. Direktoratet for Naturforvaltning, Trondhjem.

Anonymous, 2014a. Lov 24 nov 2000 nr. 82 om vassdrag og grunnvann. Pp. 2486–97, in *Norges Lover 1687–2013,* ed. by I.L. Backer and H. Bull. Det Juridiske Fakultet, Universitetet i Oslo.

Anonymous, 2014b. Lov 29 mai 1981 nr. 38 om jakt og fangst av vilt. Pp. 949–56, in *Norges Lover 1687–2013,* ed. by I.L. Backer and H. Bull. Det Juridiske Fakultet, Universitetet i Oslo.

Anonymous, 2014c. Lov 19 juni 2009 nr. 100 om forvaltning av naturens mangfold. Pp. 3663–79, in *Norges Lover 1687–2013,* ed. by I.L. Backer and H. Bull. Det Juridiske Fakultet, Universitetet i Oslo.

Berge, M., 2005. Stor salamander (*Triturus cristatus*) i Geitaknottane naturreservat. Vasshabitat- og populasjonsstudie. *Masteroppgåve i Naturvitskap, Biologi.* Universitetet i Bergen.

Chistensen, H. and Eldøy, S., 1988. Truete virveldyr i Norge. DN-rapport 2: 1988. Direktoratet for naturforvaltning, Trondhjem.

Collett, R. 1879. Bemerkninger om Norges Reptiler og Batrachier. Norske vidensk. selsk. *Christiania 1878* **3**: 1–12.

Collett, R., 1918. *Norges krybdyr og padder.* Aschehoug, Kristiania (Oslo).

Dolmen, D., 1982. Zoogeography of *Triturus vulgaris* (L.) and *T. cristatus* (Laurenti) (Amphibia) in Norway, with notes on their vulnerability. *Fauna Norvegica Serie A* **3**: 12–25.

Dolmen, D., 1993. Herptilreservat Geitaknottheiane. Forslag til verneområde for amfibier og reptiler. The Science Museum at the University of Trondheim.

Dolmen, D., 2009. De grønne froskene på Finnøy i Rogaland (Del 1–3). Zoologisk notat 2009–2. Vitenskapsmuseet, Trondhjem.

Elmberg, J., 1978. Åkergrodan. En artsöversikt samt nya rön om dess utbredning i Nord- och Mellansverige. *Fauna och Flora* **73**: 69–78.

Enger, J., 1970. Levevis og utbredelse hos spissnutet frosk *Rana arvalis,* i Fredrikstad-distriktet. *Fauna* **23**: 25–35.

Fog, K, 1995. Geografisk variation for spidssnudet frø (*Rana arvalis*). Pp. 87–9, in *Bevarelse af Danmarks padder og krybdyr,* ed H. Bringsøe and H. Graff. Nordisk Herpetologisk Forening, Køge.

Fossen, A., Strand, L.Å. and Dolmen, D., 1989. Dammer på Romerike. En zoologisk og botanisk inventering, med hovedvekt på amfibier. Fylkesmannen i Oslo and Akershus, Miljøvernavdelingen.

Gjerde, H., 1984. Amfibiefaunaen på Kongsrudmyra våren og sommeren 1983, rapport. Yngelbiologi, forekomst og fordeling i tidsrommet april til juli 1983. Meddelelse nr. 19 del A fra Kongsrudtjerngruppa.

Gjerde, L., 1989. Utbredelse og forekomst av amfibier i Nordre Øyeren naturreservat. NØBI Report 1. Nordre Øyeren Biologiske Stasjon.

Gjerde, L., 1991. Distribution and abundance of amphibians in Nordre Øyeren Nature Reserve. NØBI Report 12. Nordre Øyeren Biological Station, Lillestrøm.

Gjerde, L., 1992. Haleløse padder (Anura, Amphibia) på Kongsrudmyra, bestandstørrelse og utbredelse. Meddelelse nr. 27. Kongsrudtjerngruppa.

Gjerde, L., 1993. New color-variation in the moor frog? NØBI Newsletter **10**: 1.

Gjerde, L., 1994a. Forvaltning av amfibier og deres livsmiljø.

Gjerde, L., 1994b. Monitoring brown frogs in Nordre Øyeren 19891993. Nordre Øyeren Biological Station.

Gjerde, L., 1996. Amfibier i Øyeren-deltaet. Skoletjenesten NØBI, Nordre Øyeren Biologiske Stasjon, Lillestrøm.

Gjerde, L., 1997a. A survey of amphibian spawning sites in Rælingen municipality, Norway. NØBI Report 25. Nordre Øyeren Biological Station, Lillestrøm.

Gjerde, L., 1997b. En generell undersøkelse av potensielle ynglelokaliteter for amfibier ved Leira i Skedsmo kommune. NØBI Brief 14. Nordre Øyeren Biological Station, Lillestrøm.

Gjerde, L., 2002. En generell kartlegging av amfibier i Fet kommune. NØBI Report 26. Nordre Øyeren Biological Station, Lillestrøm.

Gjerde, L., 2008a. Kvalitetssikring av observasjoner til spissnutefrosk ved tidligere kjente yngledammer i Oslo. Kartlegging våren 2008. Naturveiledernes Oppdragsrapport 2. Norske Naturveiledere, Lillestrøm.

Gjerde, L., 2008b. Registrering av amfibier i Merkja, Nordre Øyeren naturreservat. Delrapport for konsekvensutredning på naturverdier i forbindelse med utvidelse av riksveg 22. Naturveiledernes Oppdragsrapport 4. Norske Naturveiledere, Lillestrøm.

Gjerde, L., 2009. Vurdering av avbøtende tiltak for natur langs planlagt utvidet riksveg 22 fra Borgen bro til Hovinhøgda. Naturveiledernes Oppdragsrapport 9. Norske Naturveiledere, Lillestrøm.

Gjerde, L., In prep. a. Bibliography on amphibian literature in Scandinavia. Nordre Øyeren Biological Station, Lillestrøm.

Gjerde, L., In prep. b. Monitoring amphibian populations and habitats in Nordre Øyeren Nature Reserve. NØBI Report 27. Nordre Øyeren Biological Station, Lillestrøm.

Gjerde, L., In prep. c. Neglected mitigations from expanded national road through Nordre Øyeren Nature Reserve. Nordisk Naturforvaltning.

Gjessing, J. and Ouren, T., 1983. Det Bestes Store Norges Atlas. Forlaget Det Beste, Oslo.

Gutiérrez, M. 2002. The crested newt (Triturus cristatus) in Geitaknottane Nature Reserve, Western Norway: A study on population number estimates and belly pattern variations in four ponds. Project assignment – Aitana Uria / Univ. Basque Country, Spain.

Hage, M., 1999. The northern crested newt (Triturus cristatus) in the Geitaknottane nature reserve: diet, body size and population estimates. Unpublished Cand. Scient.-thesis in biology, University of Bergen.

Hagen, Y., Norderhaug, M., and Rom, K., 1974. Truete dyrearter i Norge med Svalbard. World Wildlife Fund – Norge, Oslo.

Hansen, O.J., 1999. Dammer i Sandefjord. Registrering og vurdering. Sandefjord kommune, Sandefjord.

Heggland, A., 2010. E18 Bommestad-Sky: Bygging av erstatningsdam for amfibier. Statens vegvesen.

Hårsaker, K., Larsen, B.M. and Dervo, B.K., 2000. Overvåking av amfibier i Norge. Forslag til overvåkingsmetodikk, overvåkingsområder og deltakere i en atlasundersøkelse. NINA Oppdragsmelding 652: 1–27.

Hobæk, A., Hage, M., Solhøy, T. and Myklebust, O., 2000. Vannkvalitet og stor salamander i Geitaknottane naturreservat, Hordaland. NIVA report LNR 4261-2000.

Ingierd, D.Ø., 2006. Fiskekartbok for Oslomarka. 7th edition. Oslo fiskeadministrasjon, Oslo.

Iuell, B., 2005. Veger og dyreliv. Handbook 242 from the Road Directorate, Oslo.

Johnsen, S., 1935. Om utbredelsen av vannsalamandrene (Triton) i Norge. Naturen 59: 97–111.

Johnsson, B. and Semb-Johansson, A., 1992. Norges Dyr. Fiskene I. Cappelen, Oslo.

Kålås J.A., Viken Å. and Bakken T., 2006. Norsk Rødliste 2006. 2006 Norwegian Red List. Artsdatabanken Norge.

Kålås J.A. Viken, Å., Henriksen, S. and Skjelseth, S., 2010. The 2010 Norwegian Red List for species. Artsdatabanken, Trondhjem.

Mikalsen, H.H., 2008a: Polsk latterfrosk spreier seg. Øyposten 26 September 2008.

Mikalsen, H.H., 2008b: Latterleg frosk på Finnøy. *Stavanger Aftenblad* 4 oktober 2008.

Myklebust, O.A., 1998. Vasshabitat- og åtferdstudie av stor salamander, *Triturus cristatus* (Laurenti), i Geitaknottane Naturreservat, Ytre Hardanger. Unpublished Cand.Scient.-thesis in biology. University of Bergen.

Nilson, G. and Andrén, C., 1981. The Moor Frog, *Rana arvalis* Nilsson (Amphibia: Ranidae) on the Baltic Island Gotland, a Case of Microevolution. *Amphibia-Reptilia* **3/4**: 347–51.

Paulsen, N., 2006. Population structure of the Great Crested Newt, *Triturus cristatus*, in Geitaknottane Nature Reserve, Norway. Unpublished Cand.Scient. thesis at University of Bergen.

Paulsen, N., Tomasgård, T.E.H. and Tvedt, K., 2009. Stor salamander i Geitaknottane naturreservat, tellinger 2008. Swan-rapport 2008.

Ruud, G., 1949. Amfibiene. Pp. 21–8, in *Norges dyreliv 3*, eds B. Føyn, G. Ruud and H. Røise. Cappelen, Oslo.

Rygg, O., 1989. *Gårdsdammer i Kløfta/Borgen-området, Ullensaker kommune, Akershus. Registreringer med spesiell vekt på amfibieforekomster.* Vesong Ungdomsskole, Kløfta.

Sandlund, O.T., 1992. Biologisk mangfold i Norge. En landstudie. DN-rapport 5: 1992. Direktoratet for naturforvaltning, Trondhjem.

Schei, T. and Zimmer, F. (eds.), 1996. *Norges lover 1685–1995.* Studentutgaven. Universitetet i Oslo.

Semb-Johansson, A., 1989. The common toad – a step-child in Norwegian zoology. Fauna **42**: 174–9.

Semb-Johansson, A., 1992. Declining populations of the common toad (Bufo bufo L.) on two islands in Oslofjord, Norway. Amphibia-Reptilia **13**: 409–12.

Spikkeland, I., 1998. Dyreliv i dammer i Askim. Natur i Østfold **17**: 13–22.

Steinvåg, M.J., 2010. Forvaltingsplan for naturreservata Yddal og Geitaknottane: Naturkvalitetar, bevaringsmål og forvaltingstiltak. MVA-report 1/2010. Fylkesmannen i Hordaland, Miljøvernavdelinga, Bergen.

Strand, L.Å., 1994. Utbredelse og akvatisk habitat hos amfibier i Oslo by. Hovedoppgave i ferskvannsøkologi. Zoologisk institutt, Universitetet i Trondhjem. Upublisert.

Strand, L.Å., 2001a. Dammer på Romerike. Endringer vedrørende dammene og amfibienes bruk av disse i løpet av en 10-års periode. Fylkesmannen i Oslo og Akershus, Miljøvernavdelinga. Rapport nr. 1/2001.

Strand, L.Å., 2001b. Forslag til anleggelse av amfibietunnel i forbindelse med bygging av rundkjøring på Skoklefall, Nesodden. Statens Vegvesen Akershus.

Srand, L.Å., 2004. Forvaltningsplan for dammer og amfibier i Oslo og Akershus. Fylkesmannen i Oslo & Akershus, Oslo.

Strand, L.Å., 2015. Telling av froskeeggklaser ved Merkja, Fet, i forbindelse med anleggelse av firefelts Riksvei 22 mellom Lillestrøm og Fetsund: Resultatet for 2015. Unpublished.

Strand, L.Å., Olsen, O., Wangen, G. and Langvatn, V.A., 2009. Vårvandring i 2009 hos padde ved Litlevatnet, Volda. Statens Vegvesen.

Strand, L.Å. and Stornes, A., 2007. Amfibienes bruk av nyanlagt amfibietunnel under Tangenveien ved Skoklefall, Nesodden. Statens Vegvesen Akershus.

Strømme, A., 2005. Stor salamander (*Triturus cristatus*) i Geitaknottane naturreservat. Populasjonsestimering og fødetilgang i dammane. Unpublished Master-thesis at the University of Bergen.

Støp-Bowitz, C., 1950. Norges padder (Amphibia). *Fauna* **3**: 15–20.

Størkersen, Ø.R., 1992. Truete arter i Norge. DN-rapport 6:1992. Direktoratet for naturforvaltning, Trondhjem.

Størkersen, Ø.R., 1999. Nasjonal rødliste for truete arter i Norge 1998. DN-rapport 1999-3. Direktoratet for naturforvaltning, Trondhjem.

Wallén, C.C., 1970. *World Survey of Climatology. Volume 5.* Climates of Northern and Western Europe. Elsevier Publishing Company, Amsterdam.

62 Conservation measures and status of amphibians in Sweden

Claes Andrén

Abbreviations and acronyms used in the text and list of references

ESU   *Ecologically Significant Unit*

I. Introduction: The Scandinavian Peninsula and its amphibian fauna

During the last ice age the temperature in Europe was about 10°C lower compared to those of the present time. The Scandinavian Peninsula and surrounding areas were covered with a thick ice sheet. From 16,000 years ago the temperature started to rise and 11,000 years ago Denmark, south of Sweden, was free from ice (Fog *et al.* 1997). There are findings of amphibians (*Rana arvalis*) from 10,500 years ago; at this time the southern part of Sweden was also free from ice. For a period of 3,000 years, from 9,200 to 6,200 years ago, there was a land bridge from the continent to Sweden. This coincided with a warm climatic period, which made it possible for many heat-demanding species to migrate northward into Sweden. About 7,000 years ago the summer temperature was 1.5–2.0°C higher compared to the present time. This made it possible for species like the European pond turtle, the fire-bellied toad, treefrog, common spadefoot toad, edible frog complex, green toad and natterjack toad to spread in the southern part of Sweden. About 6,200 years ago, the sea broke through the land mass between present Denmark and Sweden and created the Baltic Sea. As a result, a gap was formed that broke the contact between Sweden and continental Europe. About 3,000 years ago, summer temperatures fell to a level approximating those of today and, during the same period, there was very limited gene flow from the Continent to Sweden. Since the ice age, 23 species of amphibians and reptiles have been resident in the Scandinavian countries and, by comparing these species and subspecies with their distribution in continental Europe, it is obvious that our present amphibians and reptiles originated in southern and southeastern Europe (Fog *et al.* 1997).

The present distribution and occurrence of different amphibian species is to a large extent explained by the size and northern position of the country. Sweden is geographically a large country, about 450,000 km^2 in area, and the distance from south to north is about 1,600 km. It covers many vegetation and climatic zones, which means that the distributions of different amphibian species vary a great deal. The northwestern parts are mountainous with elevations up to about 2,000 m. The southern part of Sweden consists mainly of agricultural areas, and 75% of the land area is forested, being the western edge of the Eurasian Taiga. There are more than 100,000 lakes of more than one hectare in size, and these lakes constitute 9% of the country's area, or 40,000 km². The number of islands changes, depending on the continuing rise of the land, but is about 270,000, including those in lakes and along the coastline. There are two major islands in the Baltic Sea.

The climate of Sweden is defined as temperate with four pronounced seasons. Despite its northern position, Sweden has large regional differences in climate as a result of the Gulfstream. In the southern part, deciduous forest dominates and in the north there are mainly coniferous forests. The climate in the north is subarctic with an average daily temperature in winter of –18°C compared to 0°C in the south. However, in summer the average daily temperature is about 17°C over most of the country (Wikipedia – Swedish Geography and Nature).

Most Swedish species of amphibians have an explosive breeding strategy as an adaptation to the comparatively cold and short summer period, during which all individuals in a population breed as early as the weather permits. The onset of breeding starts later the farther north one goes, but in the north longer days with more hours of sunshine compensate for this and speed up development.

Fog et al. (1997) discussed possible local adaptations in Scandinavian amphibians. They argued that the farther away from the central distribution a species spreads, the more limited are the conditions that the species can survive. If the spectrum of habitats and other environmental conditions is broad in the central part of the distribution, it is likely that it becomes gradually smaller when the species spreads to the north. However, if the spectrum of habitats used again becomes wider, or if the species is more specialized farther to the north, it can be perceived as of sign of local adaptation. There are some examples supporting this view. The agile frog breeds later than the common frog in central and southern Europe. In central Germany and northwards it breeds earlier than does the common frog, which may be an adaptation to cold climate. In northern Germany the agile frog seems to have populations with local adaptations. Moving farther to the north the agile frog becomes very common and is found in many different habitats; it even outcompetes other brown frogs. Tadpoles' development in the Danish populations is faster compared to the Swedish populations. The isolated Swedish populations of the agile frog are more specialized and only found in a few habitat types.

Other specializations occur in the isolated pool frog populations in the County of Uppland (north of Stockholm), where the animals are clearly darker compared to southern populations (Sjögren-Gulve 1994). In the Swedish populations of edible frogs, the triploid form dominates, which can also be an adaptation to cold climate as the tadpoles are larger at metamorphosis and give rise to larger froglets that often go directly into hibernation. The moor frog is well known to have great variation in pattern over its range, which is also true for Sweden, but there is one clear local geographical adaptation on the Baltic island of Gotland. The only brown frog occurring on Gotland is the moor frog and morphologically it has several features typical of the other two brown frogs. The hind legs are long as in the agile frog, the head is blunt and the belly pattern is dark-speckled as in the common frog, and it has a black stripe along the side of the head and body, which is unique to the Gotland populations (Nilson and Andrén 1981). Recent studies have shown that this population also is genetically unique.

II. Status of amphibian global conservation

The global decline and precarious situation for amphibians is nowadays a well-known fact to all of us. However, it was not until the first World Congress of Herpetology in Canterbury, Great Britain, in 1989, that it was evident that what scientists and conservation enthusiasts had noticed on a local or regional level proved to be a general pattern all over the world. Another important step was when Stuart et al. (2004) presented a report on the status and trends of amphibian declines worldwide, showing that amphibians were more threatened than were other vertebrate taxa. Universities, foundations and scientists in conservation biology were asked to gather resources to counter these trends.

In recent years, more than 160 species have died out, one-third of the known amphibian species are at risk of becoming extinct, and, for another quarter, our knowledge is so scarce or fragmented that they are classified as DD (data deficient) on the IUCN Red List. Most probably many of these species are also at risk of extinction. The proposed factors behind these declines are many and complex (Allentoft and O'Brian 2010; Heatwole 2013). Amphibians generally are believed to be sensitive to environmental changes and fragmentation of their habitat (Andersen *et al.* 2004), and they can be both prey and predators; they often utilize both terrestrial and aquatic habitats with different feeding strategies at different stages of their life cycle (Hels and Bushwald 2001, Kats and Ferrer 2003; Corn 2005; Cushman 2006; Gallant *et al.* 2007). The rapid spread of chyridiomycosis, caused by a pathogenic fungus, has affected amphibians more than any other known disease (Ågren and Malmsten 2008, Voyles *et al.* 2009).

III. Amphibian declines and species of special concern

Swedish amphibians benefited from the small-scale farmlands of earlier times, which included many wetlands and small, shallow ponds that were important as breeding localities. This landscape was also rich in stone cairns, shrubs, heaps of dead leaves, meadows with stands of bushes or trees, and forest edges where amphibians could find sheltered warm places and food. The small-scale landscape resulted in many suitable habitats close to each other and local adverse conditions did not play a major role because other suitable habitats could be found close by.

In order to gain more cultivable land, large-scale drainage of wetlands and ponds, and lowering the level of shallow lakes took place during the 1800s, and also continued well into the 1900s. In parts of southern Sweden about 95% of the wetland area has been drained since the mid-1800s. Because of this extensive drainage of wetlands, many important amphibian habitats were lost and thereby also many amphibian populations. After the Second World War agriculture was even more rationalized. The cohesive culture surfaces became greater, and many small biotopes such as stone cairns, small ponds and ditches were removed. At the same time, the use of chemicals in agriculture has increased. Species inhabiting shallow waters in the open landscape decreased dramatically as a result (Kjellsson *et al.* 2005).

Wetlands include a number of different damp and wet environments and ecosystems in the open or more closed landscape, inland or coastal environments, damp meadows or big lakes, moving or still water, and temporary or permanent waters. Wetlands are of major importance all over the world, which was recognized by the first international nature convention, the Ramsar Convention, which came into force in 1975. In Sweden, a wetland is normally defined as an area where most of the year water is present just beneath, at or just above the surface of the ground, and the surface of the water is covered by vegetation. About 9.3 million hectares (21% of the country) is defined as wetland. Additionally, 3.4 million hectares (8% of the country) are lakes and streams. Northern Europe differs from the rest of Europe by having much larger areas covered by wetlands. This is especially true for a type of wetland called bog. It is not only a matter of climatic differences, but there is a much greater pressure on bogs in central Europe. In Sweden there are still open, undestroyed bogs comprising an area of 4–5 million hectares, which means that we have a special responsibility for maintaining this kind of wetland (Kjellsson *et al.* 2005).

Wetlands have long been used and influenced by man for different purposes. Early settlements were often along lakes, streams or coastlines. When people started to keep cattle in stables for the winter, wetlands became even more important. A large proportion of Sweden's wetlands were used for the production of hay; periodically flooded areas and wet shorelines along streams were important grazing areas for cattle. All these biotopes were natural open areas. The management of grasslands declined dramatically during the 1900s, and now only a tiny proportion is left, often because of management focused on conservation. In an historical perspective, man used wetlands in a sustainable way, but to drain wetlands for agricultural purposes, which has occurred during the past 100–200 years has resulted in the conversion of many wetlands to other types of ecosystem.

During the second half of the 1800s, the human population in Sweden increased, and famine spread over the country. About one million people emigrated to North America during a few decades from a population of less than four million. The pressure to improve food production was very high, which resulted in a massive change of the landscape by draining wetlands and shallow lakes and by removing forests. Draining of wetlands was supported by the government. In the southern province, Scania, with the most favourable climate for thermophilic amphibians, 90% of the total area of wetlands was lost during a period of 150 years. About 2,500 lakes have either had their level lowered or were drained in Sweden. In the 1970s the agricultural area of systematic draining was estimated to be about one million hectares (Kjellsson et al. 2005). All types of wetland in the open landscape were drained.

Draining of forested areas has been going on from 1970 up to 1990, at first in only minor ways but then on an increasingly larger scale. During this period about 1.5 million hectares of forests were drained. Many new roads were built in forests; they caused hydrological disturbance and further contributed to increased drainage.

The large-scale forestry operations during the 1960s to the 1980s were often ruthless towards wetlands. Large forested areas became, from a biological viewpoint, a depleted residue of their former biodiversity. The purpose of the draining was usually either to convert open wetlands to forests or to increase timber production in wet areas. Another purpose of draining was for harvesting peat. During the 1970s and 1980s there was an increased interest in peat as fuel for district heating to households and factories (Kjellsson et al. 2005).

Even though we have very little precise information about amphibian populations during the past 150 years up to about the 1980s, our wetlands have been affected over the ages. On a national level we have lost 23% of the wetlands we had in the middle of the 1800s. Of the remaining 77% of wetlands, at least 53% is affected in different ways and only 23% of Swedish wetlands still maintain their original ecological function. However, these are mainly the coniferous forests in the harsh climate of the northern part of the country where only the more cold-adapted amphibian species thrive. In the southern third of the country, the permanent loss of wetlands is between 70–90% over the past 150 years (Kjellsson et al. 2005).

From the 1980s, however, the rate of large-scale draining of the landscape decreased markedly. Instead, many restoration projects during recent decades are bringing back as much of the lost wet ecosystems as possible. Sadly, much of the earlier draining caused permanent hydrological damage that has affected biological diversity and the available surface groundwater has decreased dramatically. The level and flow of water has been changed by regulation. Retention of water and the capacity to buffer its level has decreased. To this must be added an extensive eutrophication and acidification that has seriously affected freshwater ecosystems and added a burden of contaminants.

Owing to the major changes in the landscape described above, it is likely that all amphibian species have had a major decrease in population density during the past 100–150 years, especially in the southern half of the country. However, there has been no scientific documentation of population trends of Swedish amphibians until recently. A national monitoring program, including the common species, started in 2015. For the rarest species, mostly confined to restricted areas in

Table 62.1   Swedish Amphibians' Conservation and Protection Status 2015

Species	National protection	National Red List/ Eur. IUCN Category	Species & Habitat Directive (App.)
Lissotriton vulgaris	Yes	LC / LC	No
Triturus cristatus	Yes	LC / LC	II & IV
Bombina bombina	Yes	LC / LC	II / IV
Pelobates fuscus	Yes	VU / LC	IV
Hyla arborea	Yes	VU / LC	IV
Bufo bufo	Yes	LC / LC	No
Epidalea calamita	Yes	VU / LC	IV
Bufotes viridis	Yes	VU / LC	IV
Rana temporaria	Yes	LC / LC	V
Rana arvalis	Yes	LC / LC	IV
Rana dalmatina	Yes	VU / LC	IV
Pelophylax kl. *esculentus*	Yes	LC / LC	V
Pelophylax lessonae	Yes	VU / LC	IV

VU = Vulnerable, **LC** = Least Concern. **II** = Core habitats protected (included in Natura 2000 Network) and managed in accordance with ecological needs, **IV** = Strict protection across natural range within the European Union, **V** = Exploitation compatible with a favourable conservation status.

the south, a more thorough monitoring of population distribution, size and trends was initiated in the 1980s. It means that we have no historical information about population sizes or densities with which to compare, except for occasional localities. Many of these are not verified by museum specimens. This means that it is hard to draw conclusions about trends in populations or changes in distribution. Most information is gleaned from estimates by biologists with extensive field experience.

Sweden has 13 native species of amphibians that vary in their vulnerability to different threats, the ecological zones inhabited, their range in habitats, the size of their distributional ranges, their population densities, the protection they are afforded, and the temporal changes in these attributes (Tables 62.1–62.3). Four of these are still believed to be common: the common frog *Rana temporaria*, the moor frog *Rana arvalis*, the common toad *Bufo bufo* and the smooth newt *Lissotriton vulgaris*. They are widespread and common in the southern third of the country up to Limes Norrlandicus, a zoogeographical border marking a climatic and vegetation shift. North of this border the coniferous forest dominates the landscape. All four species occur also in lowlands and in elevated forested areas along the Baltic coast to about the Finnish border (Gislén and Kauri 1959).

The common frog *Rana temporaria* is a very competitive species. It has the widest range and is only missing on the highest mountain peaks (> 1,000 m) and on the Baltic islands. Colour and markings vary considerably within populations, but no geographic variation is observed. In mountainous northern Scandinavia, the species seems to have shorter hind legs compared to lowland specimens. The majority of the population remains within 500 m of their breeding ponds. The common frog has a wide habitat preference and seems to do well in many different landscapes, which probably explains why it is still a very common species. In the high mountainous areas larval development up to metamorphosis may take two seasons to complete (Gislén and Kauri 1959; Elmberg 1995; Fog *et al.* 1997).

The moor frog *Rana arvalis* was one of the earliest amphibian species migrating to Scandinavia and Sweden. There are records in Denmark from 10,000 years ago, meaning there has been time for local geographic adaptation. Most obvious is the variation in colour pattern and in Sweden the spotted "*maculata*" form dominates in the south and central parts of the country, while the striped "*striata*" form is more dominant in the north. Some insular populations are uni-coloured and the moor frogs on the Baltic island of Gotland constitute a unique genetic and morphological entity. There is a trend toward shorter hind legs the farther north one goes in Scandinavia (Fog *et al.* 1997; Nilson and Andrén 1981). The morphological variation and local adaptations of moor frogs in Sweden indeed needs to be investigated further. Laboratory experiments show a possible case of rapid evolutionary response to acidification (Andrén *et al.* 1989). The moor frog usually remains close to the breeding ponds and prefers areas with wet ground. Open areas with high water tables and short distances between potential breeding ponds therefore favour this species. In Denmark, Fog (1997) pointed out that the species is declining in most areas, as it is in Central Europe. The general draining of wetlands in the agricultural landscape of Sweden over the past 50–60 years has probably caused a decline in moor frog populations. However, Sweden probably still has many thriving local populations of the species.

The common toad *Bufo bufo* is protected by its poisonous skin glands, even as a tadpole. It breeds in larger permanent ponds, often with fish, as well as in brackish water. Breeding has been observed in waters with up to 7 parts per thousand of salt and surviving larvae in waters with up to 6 parts per thousand of salt (Fog *et al.* 1997). It means that the common toad normally has no amphibian competitors and few predators. They are known to have strong site fidelity and as adults they return, over long distances, to their natal pond. There are no historical data giving evidence for changes in distribution or trends in populations' development. However, being a very competitive species and doing well in many habitats and landscapes, it probably still has a stronghold in Sweden.

The smooth newt *Lissotriton vulgaris* is still a very common species. The northern populations have some unique morphological characteristics. Cyrén (1945), Gislén and Kauri (1959) and Dolmen (1978, 1981) described these populations in more detail and refer to special adaptations for surviving in a cold climate. They suggested that there should be a separate subspecies (*T. v. borealis*), but this has not been supported by later taxonomists (Griffiths 1996). An important restriction for a nocturnal species is, of course, the very short activity period due to the short summer nights in the north. These northern populations have a larger body size and a lower crest in males; neotenic populations occur along the northernmost edge of their range. Fully grown, sexually active, pigmented individuals with swollen cloacae and well-developed gills never leave the pond. Occasional specimens of neotenic great crested newts also have been found in this area. Sadly, it seems that these neotenic populations are now gone owing to release of fish into formerly fish-free small lakes. In the high elevations of montane areas larval development up to metamorphosis in the non-neotenic smooth newts may take two seasons to complete.

The great crested newt *Triturus cristatus* is more thermophilic compared to its smaller relative; it has a narrower range of habitat, uses only permanent fish-free and crayfish-free ponds with a high diversity of invertebrate prey species. The great crested newt has a continuous distribution in the southern and central parts of the country up to the boreal zone, Limes Norrlandicus. There is also an isolated relict population from the earlier post-glacial warm period along the Baltic coast in a climatically favourable area. As this is a Habitats Directive Annex II & IV species, much more attention has been given to its distribution, population size and trends in population development during the past 20–25 years. Monitoring programs have been carried out in many populations or regions as requested for Habitats Directive species with poor conservation status. In official reports from the Swedish Information Centre, the species is regarded as still having a negative trend in population density and an ongoing deterioration of its habitat (Johansson *et al.* 2005; Karlsson 2006). An interesting neotenic population occurred at the northernmost border but it is now extirpated owing to the introduction of fish. As was the case for the smooth newt, the neotenic great crested newt kept its larval characteristic of external gills, but had a swollen cloaca and was reproductively mature. These specimens were black, about 10 cm long, and completely aquatic (Cyrén 1945; Gislén and Kauri 1959; Dolmen 1978, 1981; Kupfer and Kneitz 2000; Malmgren 2001).

The remaining eight species are rarer and have restricted distributions in the southern parts of Sweden. They prefer warmer conditions and some are probably relict populations from the post-glacial warm period. Most of them are to be found on the national Red List and some are endangered. The green frog complex has, as in many European countries, a complicated taxonomic history mainly because of the poor understanding of the relations between different populations before modern genetic methods were available. One species, the fire-bellied toad, died out in the early 1960s, but has successfully been reintroduced.

The fire-bellied toad *Bombina bombina* had a rather weak population in Sweden for a very long time and was the first Swedish amphibian to become extirpated. The last observation was made in the early 1960s, and at that time almost all known localities for the species had been destroyed. The main threats to the species seemed to have been destruction of breeding habitats and the surrounding terrestrial areas, over-application of fertilizers, planting of coniferous forests, and introduction of fish and crayfish. It inhabits shallow wetlands but is also found in permanent and fish-free ponds exposed to the sun. Preferred breeding ponds have dense coverage of floating and submersed macrophytes. In 1982 an extensive and successful re-introduction project was initiated (Nilsson 1954; Madej 1973, Berglund 1976, 2007; Andrén and Nilson 1986, 2000; Krone and Künel 1996; Fog *et al.* 1997; Andrén and Berglund 2002, Briggs and Damm 2004).

The spadefoot toad *Pelobates fuscus* has a limited distribution in the southern part of the province of Scania. It is restricted to sandy soils and is vulnerable to extirpation because its ability to recolonize is poor, owing to relatively low mobility and high site-fidelity. The low dispersal rates of adults between ponds may cause local extirpation in the long term since many populations are dependent on immigration by froglets from surrounding populations. The spadefoot toad lives in metapopulations and has limited reproductive success in about 50% of the breeding ponds containing high concentrations of nutrients. This suggests that changes in water chemistry in

Table 62.2　Trends in Swedish Amphibian Habitats Directive Species 2013

Species	Alpine zone 1 2 3 4 5	Boreal zone 1 2 3 4 5	Continental zone 1 2 3 4 5	Distribution area (km²)	Used habitat area (km²)	Total population size (individuals)
Triturus cristatus		o – – – –	o – – – –	192.900	110	325.000
Bombina bombina			o o o o o	2.300	240	20.000
Pelobates fuscus			o o o o o	3.300	50	5.000
Hyla arborea			o o o o o	3.100	150	62.000
Epidalea calamita		o o o o o	o o + o o	11.600	95	38.400
Bufotes viridis			+ + + + +	2.900	4	1.300
Rana temporaria	o o o o o	o o o o o	o o o o o	497.600	100.000	442 mn
Rana arvalis	o o o o o	o o o o o	o o o o o	442.900	89.000	405 mn
Rana dalmatina		o – o o o	o o o o o	11.700	530	32.000
Pelophylax kl. *esculentus*		o o o o o	o o o o o	3.700	16	25.200
Pelophylax lessonae		o o o o o		2.100	60	12.000

Data from the Species Conservation (ArtDatabanken), Species and Habitats – Conservation Status in Sweden 2013. 1 = Distribution area, 2 = Population size, 3 = Habitat quality, 4 = Prospects, 5 = Overall assessment. Comparison between 2007 and 2013: o = No change, – = Negative trend, + = Positive trend. Empty zone = No records of the species. mn = Millions.

association with agriculture and animal husbandry may influence reproductive success and local population densities. The spadefoot toad has decreased dramatically since 1959, from about 100,000 individuals in more than 400 breeding ponds to fewer than 1,000 specimens in 55 ponds in the mid-1990s. In the national Red List 1996, it was categorized as Critically Endangered (CR). At that time a massive restoration program began to improve the situation for this species (Berglund 1976; Axelsson *et al.* 1997, 1998; Hels and Buchwald 2001; Edenhamn and Sjögren-Gulve 2002; Hels 2002; Oritz *et al.* 2004; Releya 2005; Nyström *et al.* 2008a).

The natterjack toad *Epidalea calamita* has a limited distribution along the southern and south-western coastal area, mainly in the continental biogeographic zone, but just entering the boreal zone at its northwestern distributional limit. The natterjack toad lives primarily in open and dry environments on sandy or rocky soil with low or scarce vegetation. Typical biotopes exist on rocky islands, in coastal meadows, heathland, and sand pits. These biotopes are in many cases dependent upon grazing. Diminished grazing, tree plantations, drainage and increased exploitation have caused a substantial decline, especially in sandy areas. The decline has been worse in the inland sandy sites in Scania where many local populations have died out, or have nearly done so. Coastal rocky areas and small islands have managed better. The natterjack toad has a very prolonged reproductive period in Sweden compared with other species, which is especially true in areas with shallow rock pools used as breeding sites on small islands with bare rocks. These waters are ephemeral and often dry out before the tadpoles metamorphose. An interesting adaptation to these unpredictable conditions is that females within the same population have different strategies: some deposit eggs early and others late but over an extended period. Some even deposit part of the egg mass early and the rest later. Depending on the weather each year, different strategies pay off better than others. This species has been monitored yearly for the past 20–25 years (Andrén and Nilson 1979, 1985a, 1985b; Beebee 1983, 1985; Silverin and Andrén 1992; Sinsch 1998; Svensk naturförvaltning 2008; Allentoft *et al.* 2009; Rogell 2009).

The green toad *Bufotes viridis* is Sweden's rarest amphibian. The Swedish populations belong to a subgroup derived from Turkey and, apart from Sweden, this species also occurs in Denmark, northern Germany, Poland and the Baltic states. In Sweden the species is restricted to steppe-like habitats such as coastal pastures, bare rocks and sparsely vegetated stone quarries. It also breeds in brackish seawater along the southern coast. The disappearance of suitable habitats through overgrowth, drainage and lowering of groundwater explains the sharp decline beginning in the 1950s and continuing for the following 50 years. In the mid-1990s, only three breeding localities with altogether about 500 individuals remained. At this time, major conservation efforts started to save the green toad as it was on the brink of extirpation (Berglund 1976; Fog *et al.* 1997; Oritz *et al.* 2004; Wirén 2006; Nyström and Stenberg 2008a).

The treefrog *Hyla arborea* reproduces in permanent ponds, usually of less than one hectare in size, usually located in pastureland with deciduous trees. It breeds during a period of four to five weeks starting in early May and metamorphosis takes place from July to September, depending on the weather. After the breeding period, the adults are found in bushes and tall herbs and often in the canopy high up in trees. Froglets are found in thorny bushes. A mosaic of grazed pastures and free-standing shrubbery is favourable for the treefrog. The species is vulnerable to draining, introduction of fish and crayfish, and afforestation of pastures. Historically the species is known from more than 850 localities in the southern part of Scania province but, owing to habitat destruction and deterioration, fewer than 180 localities with about 2,500 males remained in 1989 and the species was classified as Critically Endangered (CR). In the early 1990s a massive habitat restoration program was started (Berglund 1976; Fog *et al.* 1997; Andersen *et al.* 2004; Briggs and Damm 2004).

The agile frog *Rana dalmatina* has its main distribution in central Europe with a few relict populations in northern Germany, and in the southeastern parts of Denmark and Sweden. The agile frog breeds very early, compared to other Swedish amphibians, in climatically favourable areas, some years already by January or February, but normally in mid-March. Breeding sites are normally shallow fens or ponds with areas of open water. Populations are found in or near deciduous forest. It is a very mobile species with good dispersal ability. The populations fluctuate very much and local extirpations are common, explained by weather and local events, but recolonization usually follows within a few years if there are surrounding sites. Common threats are cessation of grazing by cattle in forested areas, spruce plantations, introduction of fish or crayfish, and pesticides (Ahlén *et al.* 1995; Fog *et al.* 1997; Ahlén 2013).

The edible frog *Pelophylax* kl. *esculentus* has a small range in the southwestern corner of the southern province, Scania. There are also four very small and isolated local populations along the eastern side of southern Sweden up to Stockholm. It is not clear if these are relict populations from the earlier post-glacial warm period or old introductions by humans. In areas where habitat and climate is favourable, the edible frog is very competitive. One example of this is a recent illegal introduction on the Swedish western coast at Gothenburg about 20 years ago. The species is now spreading quickly and threatens local populations of brown frogs. Earlier, there was an interesting triploid population of edible frogs in the eastern part of Scania, but it died out about 1990 (Ahlén *et al.* 1995).

The pool frog *Pelophylax lessonae* has about 60 local populations around breeding ponds in a coastal area on the eastern side of central Sweden in County Uppland and north of the city of Uppsala. It is a very isolated group of small populations far away from those of other pool frogs. They are not only isolated from the rest of the European populations, but are also the most northerly ones. Recently, pool frogs have been discovered in a few small ponds on the Swedish east coast where they constitute about 10% of the green frog population, the remaining being edible frogs. The pool frogs are dependent upon warm and sunny shallow ponds for their reproduction, which starts in late May to early June. Metamorphosis usually takes place in late August or in September, but during cold summers reproduction might fail. Genetic analysis has shown that the Swedish pool frogs must be regarded as an ESU adapted to a harsher climate and with shorter growth period than are pool frogs in other areas (Sjögren *et al.* 1988; Sjögren 1991; Sjögren-Gulve 1994; Tegelström and Sjögren-Gulve 2004; Nilson and Pröjts 2007; Arioli *et al.* 2010; Oriazola *et al.* 2010; Nilsson 2013).

IV. Conservation measures and monitoring programmes

All amphibian species in Sweden are protected by national law from being intentionally moved, harmed or killed. However, they are not protected from habitat developments such as the building of a house or road. Ponds with open water are protected and cannot be drained nowadays without special permission. A more powerful protection is accorded to species listed in the Habitats Directive when implemented by Swedish national law. If at the same time they are listed in the national Swedish Red List, the protection is even greater.

In the 1980s, attention to the troublesome situation for many amphibian species was recognized by national and regional conservation authorities. Action plans for eight species have gradually been implemented. The conservation status now is much better for many species. However, it should be kept in mind that even if we can proudly show improvements in conservation status, such as increasing populations, for almost all amphibian species, it starts from a very low level compared to former times (Nyström and Stenberg 2008a; Andrén and Hallengen 2012).

The great crested newt *Triturus cristatus* is widespread geographically and in 2005 it was downgraded from NT (Near Threatened) to LC (Least Concern) after the outcome of surveys at local or regional scales in many parts of southern and central Sweden. However, in some counties the species could not be found at several areas from which it was previously known. Over 4,000 ponds, pools and small lakes have been surveyed and reported since 1990, and approximately 900 sites with great crested newts have been identified. Some of these are probably not used for reproduction. Conservation work for the great crested newt and its habitats and ponds is still in a rudimentary state. However, well-established, standardized methodologies adapted for Swedish conditions are available both for surveying and for monitoring. As the species is a strongly protected by the Habitats Directive, compensatory measures are always taken when the species or its habitat are at risk of exploitation. This means that the great crested newt has a better long-term outlook than do most other amphibians. A massive programme for digging and restoring ponds for endangered species in the southern parts of the country has been positive for the great crested newt. The present action plan includes information measures such as advice to land owners, inventories and monitoring, as well as efforts to manage, restore and create habitats in landscapes with functional ponds (Langton *et al.* 2001; Johansson *et al.* 2005; Karlsson 2006; Malmgren 2007; ArtDatabanken 2015).

The fire-bellied toad *Bombina bombina* is known historically from 20 localities in southern and western Scania. Only one of these local populations has remained relatively stable during the 1900s and it survived up to about 1960, after which the species was regarded as extirpated from the country. In earlier times, the fire-bellied toad was introduced onto many estates, which makes it more difficult to understand and describe the natural distribution and the status of populations of this species. In the mid-1970s, illegal re-introductions were made using hybrid animals (*B. bombina* x *B. variegata*) from Central Europe and these animals most probably died out after a few years. The Swedish National Conservation Authority authorized an attempt to bring back the fire-bellied toad in Sweden. During a twenty-year period starting 1982 the fire-bellied toad was re-introduced into eight different restored areas in Scania. Over three years, eggs were collected each time from > 10 different females from different Danish populations. The resulting young were reared in captivity up to a few weeks after metamorphosis and then released in nature. Part of the Danish material was used as a captive population to produce new toads for release for the next 17 years. In 1993, fire-bellied toads were reported from 15 different breeding ponds and the population was estimated to be about 300 individuals. Owing to long periods of drought in several of the new ponds, the population remained low for another 6–7 years. During the 1980s and 1990s, hundreds of new pond clusters for amphibians were constructed in Scania. The fire-bellied toad lives in metapopulations and seems to disperse well. From the year 2000 and several years onwards the Swedish fire-bellied toad population doubled each year. In 2005 it was downgraded on the national Red List to NT (Near Threatened) and from 2010 it has been removed from the Red List and remains strong. In 2015, the population is estimated to be at least 20,000 individuals in more than 300 breeding ponds (Andrén and Nilson 1986; Andrén and Berglund 2002; Berglund 2007; Stenberg and Nyström 2010; Andrén and Hallengren 2012; ArtDatabanken 2015).

The spadefoot toad *Pelobates fuscus* diminished rapidly from the 1960s through the 1990s, probably owing to intensification of agriculture and extensive elimination of ponds from the landscape. During a 30-year period, Sweden lost about 99% of its known population of spadefoot toads. In the late 1990s, between 400 and 850 individuals were observed in 70 different ponds. A new Action plan in 2001 stated that a first goal should be 75 viable breeding populations with a total of at least 1,500 calling males. A combination of habitat restoration and captive-bred toads being released in nature resulted in the population markedly increasing during the next ten years. The long-term goal is to have 150 ponds with 4,000 breeding males in eight different regions. In 2015, the present population is 5,000 adults in an area of 3,300 km^2 and a used area of 50 km^2 (Berglund 1976, 1998; Nyström *et al.* 2002; Nyström and Stenberg 2008b; Andrén and Hallengren 2012; ArtDatanaken 2015).

During the past 50 years, the treefrog *Hyla arborea* has been reported from at least 850 localities in the southern province, Scania. However, owing to the deterioration and destruction of habitats no more than 180 localities and about 2,500 calling males were recorded in 1989. During the 1990s, hundreds of ponds were constructed or restored and in combination with favourable weather this helped the treefrog population partly to recover. In the year 2000 about 14,000 calling males were recorded at 410 localities in Scania. From being classified as Critically Endangered in 1977 and Vulnerable in the 1990s, it was downgraded to Near Threatened in the year 2000. Owing to its increase since then, it was removed from the Red List shortly after the year 2000 and has remained as Least Concern up to the present. The long-term objective is that through adequate management of ponds, pastureland and deciduous forests, the treefrog shall attain sufficient numbers to recolonize northeastern Scania and part of the adjacent County Blekinge, where it formerly occurred. In 2015 the population was estimated to be 62,000 individuals in an area of 3,100 km^2 (Berglund 1976; Andersen *et al.* 2004; Nyström and Stenberg 2008a; Andrén and Hallengren 2012; ArtDatabanken 2015).

The natterjack toad *Epidalea calamita* is confined to coastal areas in the southern and southwestern parts of Sweden. Along the west coast, the species has been confirmed to occur on more than 40 small rocky islands since 1975. In this archipelago there are about 150 islands with rock pools suitable for breeding that may constitute potential habitat for a dynamic metapopulation of natterjack toads stochastically colonizing and becoming extinct on the various islands. The west-coast population is regarded to be relatively stable. More to the south, the natterjack toad has suffered from a strong decline during the past 50 years and most inland localities no longer contain this species. In the year 2000, the province of Halland had three or four small populations remaining. In the same year, in Scania, the species was found at 30 localities, compared to at least 200 localities 40 years earlier. In the province of Blekinge, less than ten small populations were recorded in the year 2000. Despite the restoration of habitats in the sandy localities in southern Sweden, these populations are under permanent threat and have declined in both size and number. At present, the total Swedish population of the natterjack toad is classified as Vulnerable. In 2015, the population was estimated to be 38,400 individuals in an area of 11,600 km^2 (Andrén and Nilson 1986; Briggs 2004; Pröjts 2008, 2012; Svensk Naturförvaltning AB 2008; Allentoft *et al.* 2009; Rogell 2009; Andrén and Hallengren 2012; ArtDatabanken 2015).

The Swedish occurrence of the green toad *Bufotes viridis* earlier included more than 50 known localities, mostly coastal areas along the southern and southeastern coast of the country. During the past 50–60 years, however, local populations have been lost continuously. In the year 2000, the total population was estimated to be not more than 400 individuals and with only three populations having more than 50 adults. The reason for the species' decline is complex, which is true for many amphibians, and it reflects the continuous change in the cultural landscape over the past hundred years. General reasons for decline are eutrophication of breeding waters and deterioration of surrounding terrestrial feeding areas. Since 1995 a habitat restoration, reinforcement and reintroduction program has been implemented within the known historical range of the species. Captive breeding on a large scale with reintroduction of larvae, newly metamorphosed toads and subadults has been carried out in restored localities. For several years the restocking attempts did not give desirable results and there was only a slight population increase. In 2008, the total number of adults was estimated to be 1,200 individuals. However, the last 5–6 years of restoration and restocking

Table 62.3 Temporal Trends in Swedish Amphibians' Conservation Status (National Red List)

Species	1977	1988	1992	1996	2000	2005	2010	2015
Lissotriton vulgaris	+	+	+	+	+	+	+	+
Triturus cristatus	?	?	–	–	+	+	+	+
Bombina bombina	RE	– – – –	– – – –	– – –	– – –	–	+	+
Pelobates fuscus	– – – –	– – – –	– – – –	– – – –	– – –	–	– –	– –
Hyla arborea	– – –	– – –	– –	– –	+	+	+	+
Bufo bufo	+	+	+	+	+	+	+	+
Epidalea calamita	– –	– –	– –	– –	– –	– –	– –	– –
Bufotes viridis	– – – –	– – – –	– – – –	– – – –	– – – –	– – – –	– – – –	– –
Rana temporaria	+	+	+	+	+	+	+	+
Rana arvalis	+	+	+	+	+	+	+	+
Rana dalmatina	– –	– –	– –	– –	– –	– –	– –	– –
Pelophylax kl. esculentus	–	–	+	+	+	+	+	+
Pelophylax lessonae	– –	– –	– –	– –	– –	– –	– –	– –

RE= Regionally extinct, – – – – – = **CR** (Critically Endangered), – – – – = **EN** (Endangered), – – – = **VU** (Vulnerable), – – = **NT** (Near Threatened), += **LC** (Least Concern)

have led to a marked population increase in a few localities. In the Red List of 2015 the species was downgraded to Vulnerable owing to the predicted future survival of the population. We estimate the present population to be 1,300 adults in an area of 2,900 km^2 (but the actual habitat used is only 4 km^2 at a few localities) (Briggs 2004; Wirén 2006, 2010; Andrén and Hallengren 2012; ArtDatabanken 2015).

An action plan for the agile frog *Rana dalmatina* has been established to secure its long-term survival in Sweden and to mitigate the immediate threats to its populations. Information on the ecology and status of the species is presented, the most important threats are described and measures to improve the situation are suggested. The agile frog has good dispersal ability, but spruce plantations and cultivated fields can act as barriers. Therefore, corridors with suitable habitat will stabilize populations that can otherwise fluctuate markedly. Extirpations are normal but recolonizations usually follow within a few years from surrounding areas. A number of threats have been identified, e.g. cessation of grazing, spruce plantations, dumping of waste material, introduction of fish or crayfish, urbanization and pesticides. A vision for the future is an improvement of the status that allows the agile frog to leave the Red List. A short-term objective is to stop the ongoing reduction of the distributional area and facilitate connectivity between different populations. In the Red List of 2015 the species is graded as VU (Vulnerable). We estimate the present population to be 32,000 adults in an area of 11,700 km^2 and with an area actually used of 530 km^2 (Ahlén *et al* 1995; Andrén and Hallengren 2012; Ahlén 2013; ArtDatanaken 2015).

The action plan for the pool frog *Pelophylax lessonae* began in the year 2000 in order to secure the species' survival in Sweden and to counteract the factors negatively influencing its population size and its distribution. The Swedish populations are not only isolated from the rest of the European populations, but they are also the most northerly ones. The Upplandic populations seems to be more locally adapted to a harsher climate with a short period for activity and growth in summer compared to the conditions experienced by pool frogs from Poland and Latvia. These adaptations, in combination with isolation from other pool frog populations, have caused the Upplandic Pool frogs to be regarded as an Evolutionary Significant Unit (ESU). Hitherto, actions have focused on monitoring and analysis to improve the knowledge about these unique populations of pool frogs and to restore breeding localities and surrounding terrestrial habitat. Future actions will be based

on these successful methods. The species is still regarded as Vulnerable in Sweden with an estimated total population of 12,000 specimens in an area of 2,100 km^2 (Sjögren-Gulve 1994; Fog *et al.*
1997; Andersen *et al.* 2004; Cushman 2006; Gallant *et al.* 2007; Nilsson and Pröjts 2007; Arioli *et al.*
2010; Oriazaola *et al.* 2010; Andrén and Hallengren 2012; Nilsson 2013; Lindgren *et al.* 2014;
ArtDatabanken 2015).

V. Conclusions

Compared to other species of plants and animals in Sweden, as well as amphibians globally, the
population development in frogs, toads and newts in Sweden has been positive. However, it should
be kept in mind that this trend reflects the past 20–30 years and is due to a purposeful investment
in creating new habitats for endangered amphibians. Even if most species have a better conservation status today and some have been removed from the national Red List, we have lost a large
proportion compared to 50 or 100 years ago. This is mostly because of the radical change in the
landscape, which is true for all European countries. Still, Sweden has proven that purposeful and
determined restoration achieves results.

VI. Acknowledgements

I would like thank the following persons for important contributions to amphibian conservation
in Sweden: Ingemar Ahlén, Sven-Åke Berglind, Boris Berglund, Johan Elmberg, Kåre Fog, Susanne
Forslund, Anders Hallengren, Jon Loman, Jan Malmgren, Göran Nilson, Per Nyström, Christer
Persson, Jan Pröjts, Björn Rogell, Per Sjögren-Gulve, Marika Stenberg, and Mats Wirén.

VII. References

Ågren, E. and Malmsten, J. 2008. Jordens groddjur hotas av infectionssjukdomar. (Amphibians globally are threatened by infectious diseases). *Fauna och flora* **103**: 2–7.

Ahlén, I. 2013. Åtgärdsprogram för långbensgroda 2013–2017. (Action plan for the Agile frog 2013–2017). Rapport 6586, Naturvårdsverket.

Ahlén, I., Andrén, C. and Nilson, G., 1995. *Sveriges grodor, ödlor och ormar.* (Swedish amphibians, lizards and snakes). Naturskyddsföreningen.

Allentoft, M.E., Siegismund, H.R., Briggs, L. and Andersen, L.W. 2009. Microsatellite analysis of thr Natterjack toad (*Bufo calamita*) in Denmark: populations are islands in a fragmented landscape. *Conservation Genetics* **10**: 15–28.

Allentoft, M.E. and O'Brian, J. 2010. Amphibian declines, loss of genetic diversity and fitness: A review. *Diversity*, 2.

Andersen, L.W., Fog, K. and Damgaard, C., 2004. Habitat fragmentation causes bottlenecks and inbreeding in the European tree frog (*Hyla arborea*). *Proc. Roy. Soc. London Ser. B* **271**: 1293–1302.

Andrén, C. and Berglund, B. 2002. Lägesrapport klockgroda. (The present status of the Fire-bellied toad, 2002). Rapport till Naturvårdsverket.

Andrén, C. and Hallengren, A., 2012. Framgångar för Sveriges groddjur, pp. 19–21. (Success for Sweden's amphibians). In *Skog och mark – om tillståndet i svensk landmiljö.* Naturvårdsverket.

Andrén, C., Mården, M. and Nilson, G., 1989. Tolerance to low pH in a population of Moor frogs Rana arvalis Nilsson, from an acid and a neutral environment – a possible case of rapid evolutionary response to acidification. *Oikos* **56**: 215–23.

Andrén, C. and Nilson, G., 1979. Om stinkpaddans Bufo calamita utbredning och ekologi på den svenska västkusten. (Distribution and ecology of the Natterjack toad, *Bufo calamita*, at the Swedish west coast). *Fauna och flora* **74**: 121–32.

Andrén, C. and Nilson, G., 1985a. Breeding pool characteristics and reproduction in an island population of Natterjack toads, *Bufo calamita* (Laur.), at the Swedish west coast. *Amphibia-Reptilia* **6**: 137–42.

Andrén, C. and Nilson, G., 1985b. Habitat and other environmental characteristics of Natterjack toad (*Bufo calamita* Laur.) in Sweden. *Brit. J. Herpetol.* **6**: 419–24.

Andrén, C. and Nilson, G., 1986. Klockgrodans, Bombina bombina (L.), biologi och nuvarande status vid sin nordvästliga utbredningsgräns. (The Fire-bellied toad's, *Bombina bombina* (L.), biology and present status at its northwestern distribution border). *Fauna och flora* **81**: 1–16.

Andrén, C. and Nilson, G., 2000. Åtgärdsprogram för bevarande av klockgroda (Bombina bombina). (Action plan for the conservation of the Fire-bellied toad (*Bombina bombina*). Naturvårdsverket. (The Swedish Environmental Protection Agency.)

Arioli, M., Jakob, C. and Reyer, H.U., 2010. Genetic diversity in water frog esculentus) varies with population structure and geographic location. *Molecular Ecology* **19 (9)**: 1814–28.

ArtDatabanken, 2015. *Rödlistade arter I Sverige 2015.* (Red listed species in Sweden 2015). ArtDatabanken SLU, Uppsala.

Axelsson, E., Nyström, P., Sidenmark, J. and Brönmark, C., 1997. Crayfish predation on amphibian eggs and larvae. *Amphibia-Reptilia* **18**: 217–28.

Beebee, T.J.C., 1983. *The Natterjack Toad.* Oxford University Press, Oxford, UK.

Berglund, B., 1976. Inventering av Skånes sällsynta groddjur. (Survey of Scania's rare amphibians). Statens Naturvårdsverk PM 765.

Berglund, B., 1998. Projekt lökgroda 1993–1996. (Project Spadefoot toad 1993–1996). Meddelande nr 98:9, Länsstyrelsen i Skåne län.

Berglund, B., 2007. Projekt klockgroda – historic och status fram till 2005. (Project Fire-bellied toad – history and status up to 2005). Rapport 2007: 200, Länsstyrelsen I Skåne län.

Briggs, L., 2004. Restoration of breeding sites for threatened toads on coastal meadows. In: Coastal meadow management, LIFE00NAT/EE/7083.

Briggs, L. and Damm, N., 2004. Effects of pesticides on *Bombina bombina* in natural pond ecosystems. Pesticides Research No 85. Miljöministeriet, Danmark.

Corn, P.S., 2005. Climate change and amphibians. *Anim. Biodivers. Conserv.* **28**: 59–67.

Cushman, S.A., 2006. Effects of habitat loss and fragmentation on amphibians: a review and prospectus. *Biol. Conserv.* **128**: 231–40.

Cyrén, O., 1945. Fynd av större vattenödlan vid Stensele. (Observations of the Great Crested Newt at Stensele). *Fauna och flora* **40**: 238.

Dolmen, D., 1978. De neotene salamandrarne (skrattaborrerne) ved Stensele. (The neotenic salamander at Stensele). *Fauna och flora* **73**: 161–200.

Dolmen, D., 1981. Ett nytt "skrattaborrtjärn" i Lappland. (A new neotenic salamander pond in Lappponia). *Fauna och flora* **76**: 89–92.

Edenhamn, P. and Sjögren-Gulve, P., 2002. Åtgärdsprogram för bevarande av lökgroda (Pelobates fuscus). (Action plan for the conservation of the Spadefoot toad (*Pelobates fuscus*). Naturvårdsverket.

Eide, W. (ed.), 2014. Arter och naturtyper i Habitatdirektivet – bevarandestatus i Sverige 2013. (Species and habitat types in the Habitat Directive – conservation status in Sweden 2013). ArtDatabanken SLU, Uppsla.

Elmberg, J., 1995. Grod- och kräldjurens utbredning i Norrland. (Distribution of amphibians and reptiles in north of Sweden). *Natur i Norr* **14**: 57–82.

Fog, K., Schmedes, A. and Rosenörn de lasson, D., 1997. *Nordens padder og krypdyr*. (Scandinavian amphibians and reptiles). Gad/Copenhagen.

Gallant, A.L., Klaver, R.W., Casper, G.S., and Lannoo, M.J., 2007. Global rates of habitat loss and implications for amphibian conservation. *Copeia* **4**: 967–979.

Gislén, T. and Kauri, H., 1959. Zoogeography of the Swedish amphibians and reptiles, with notes on their growth and ecology. *Acta vertebratica* **1**: 191–397.

Heatwole, H., 2013. Worldwide decline and extinction of amphibians. Chapter 13 (pp. 259–278) in *The Balance of Nature and Human Impact*, ed. by K. Rohde. Cambridge University Press, Cambridge.

Hels, T., 2002. Population dynamics in a Danish metapopulation of Spadefoot toads *Pelobates fuscus*. *Ecography* **25**: 303–13.

Hels, T. and Buchwald, E., 2001. The effect of road kills on amphibian populations. *Biol. Conserv.*, **99**: 331–40.

Johansson, N., Mernelius, P. and Apelqvist, M., 2005. Större vattensalamander (Triturus cristatus) I Jönköpings län – en sammanställning av inventeringar 2004–2005. (Great Crested Newt (Triturus cristatus) in Jönköping County – a compilation of surveys 2004–2005). Meddelande 2005:43. Länsstyrelsen i Jönköpings län, Jönköping.

Karlsson, T., 2006. Större vattensalamander (Triturus cristatus) I Östergötland: Sammanställningar av inventeringar 1994–2005 och övriga fynd I Östergötlands län. (Great Crested Newt (*Triturus cristatus*) in the County Östergötland: Compilation of surveys 1994–2005 and other findings in the County of Östergötland). Rapport 2006:4. Länsstyrelsen i Östergötlands län, Linköping.

Kats, L.B. and Ferrer, R.P., 2003. Alien predators and amphibian declines: review of two decades of science and transition to conservation. *Divers. Distrib.*, **9**: 99–110.

Kjellsson, A., Löfroth, M., Pettersson, Å. and von Essen, C., 2005. Våtmarksstrategi för Sverige. (Swedish Wetland Strategy). Rapport från Världsnaturfonden WWF, Sveriges Ornitologiska Förening SOF, Svensk Våtmarksfond SF and Svenska Jägareförbundet SJF.

Krone, A. und Künel, K.-D., 1996. *Die Rotbauchunke (Bombina bombina) – Ökologie und Bestandssituation*. Rana 1, 1–133. Natur & Text, Berlin.

Kupfer, A. and Kneitz, S., 2000. Population ecology of the Great crested newt (Triturus cristatus) in an agriculture landscape: dynamics, pond fidelity and dispersal. *Herpetological Journal* **10**: 165–72.

Langton, T. Beckett, C. and Foster, J., 2001. *Great crested newt conservation handbook*. Froglife, Suffolk.

Lindgren, B., Nilsson, J. and Söderman, F., 2014. Åtgärdsprogram för gölgroda 2014–2019. (Action plan for the Pool frog 2014–2019). Rapport 6631, Naturvårdsverket.

Madej, Z., 1973. Ecology of European bellied toads (*Bombina*) (Oken, 1816). *Przgl. Zool.* **17**, 200–4.

Malmgren, J.C., 2001. *Evolutionary ecology of newts. Doktorsavhandling*. Örebro Studies in Biology 1. Örebro University, Örebro.

Malmgren, J., 2007. Åtgärdsprogram för bevarande av större vattensalamander Triturus cristatus och dess livsmilöer 2006–2010. (Action plan for the conservation of Great crested newt and its habitats 2006–2010). Naturvårdsverket, Rapport 5636.

Niesel, J. and Berglind, S-Å., 2003. Habitat och hotsituation för större vattensalamander (Triturus cristatus). Sammanställning och utvärdering av inventeringar i Värmlands län 1991–2003. (Habitat and threat to the Great crested newt (Triturus cristatus). A compilation and evaluation of surveys in the County of Värmland 1991–2003.) Karlstad: Länsstyrelsen i Värmlands län. Miljöenheten (rapport nr 2003:16).

Nilsson, O.H.A., 1954. On the larval development and ecological conditions governing the distribution of the Fire-bellied toad, *Bombina bombina* (L.), in Scania. Kungl. Fysiogr. Sällsk. Handl. N.F. 25, 10, 124. Lund.

Nilsson, J., 2013. 2009 års inventering av gölgroda (Rana lessonae) i Norduppland. (The 2009 monitoring of the Pool frog (*Rana lessonae*) in the northern part of the County Uppland.) *Meddelandeserien* 2013:02. Länsstyrelsen i Uppsala län.

Nilsson, J. and Pröjts, J., 2007. 2005 års inventering av gölgroda längs Nordupplands kustband samt utvärdering av gölgrodans åtgärdsprogram. (The 2005 monitoring of the Pool frog along the northern coastline in the County Uppland and evaluation of the Pool frog Action plan). *Meddelandeserien* 2007:1. Länsstyrelsen i Upplands län.

Nyström, P., Birkedal, L., Dahlberg, C. and Brönemark, C., 2002. The declining Spadefoot toad Pelobates fuscus: calling site choice and conservation. *Ecography* **25**: 488–98.

Nyström, P. and Stenberg, M., 2008a. Forskningsresultat och slutsatser för bevarandearbetet med hotade amfibier i Skåne – en litteraturgenomgång. (Research results and conclusions for the conservation work with threatened amphibians in Scania – a literature review). Rapport 2008:55. Länsstyrelsen i Skåne län.

Nyström, P. and Stenberg, M., 2008b. Åtgärdsprogram för lökgroda 2008–2011. (Action plan for th Spadefoot toad 2008–2011). Rapport 5826, Naturvårdsverket.

Oriazaola, G., Quintela, M. and Laurila, A., 2010. Climatic adaptation in an isolated and getetically impoverished amphibian population. *Ecography* **33**: 730–7.

Oritz, L.-H., Marco, A., Saiz, N. and Lizana, M., 2004. Impact of ammonium nitrate on growth and survival of six European amphibians. *Archives of Environmental Contamination and Toxicology* **47**: 234–9.

Podloucky, R. and Manzke, U., 2003. Verbreitung, Öklogie und Schutz der Wechselkröte (Bufo viridis). *Mertensiella* **14**: 1–327.

Pröjts, J., 2008. Strandpaddan i Sverige 2008. Utvärdering av åtgärdsprogrammet. (The Natterjack toad in Sweden 2008. Evaluation of the Action plan). Rapport 2009: 20. Länsstyrelse I Skåne län.

Pröjts, J., 2012. Åtgärdsprogram för strandpadda 2013–2017. (Action plan for the Natterjack toad 2013–2017). Rapport 6539, Naturvårdsverket.

Relyea, R.A., 2005. The impact of insecticides and herbicides on the biodiversity and productivity of aquatic communities. *Ecological Applications* **15**: 618–27.

Rogell, B., 2009. *Genetic variation and local adaptation in peripheral populations of toads*. Acta Universitatis Upsaliensis. Uppsala.

Silverin, B. and Andrén, C., 1992. The ovarian cycle in the Natterjack toad, *Bufo calamita*, and its relation to breeding behaviour. *Amphibia-Reptilia* **13**: 177–92.

Sinsch, U., 1998. *Biologie und Ökologie der Kreuzkröte*. Laurenti Verlag, Bochum.

Sjögren, P., 1991. Genetic variation in relation to demography of peripheral pool frog populations (*Rana lessonae*). *Evolutionary Ecology* **5**: 248–71.

Sjögren, P., Elmberg, J. and Berglind, S.-Å., 1988. Thermal preference in the pool frog Rana lessonae: impact on the reproductive behavior of a northern fringe population. *Holartic Ecology* **11**: 178–84.

Sjögren-Gulve, P., 1994. Distribution and extinction patterns within a northern metapopulation of the pool frog, Rana lessonae. *Ecology* **75**: 1357–67.

Stenberg, M. and Nyström, P., 2010. Åtgärdsprogram för bevarande av klockgroda Bombina bombina 2010–2014. (Action plan for conservation of Fire-bellied toad *Bombina bombina* 2010–2014). Naturvårdsverket, Rapport 6363.

Stuart, S.N., Chanson, J.S., Cox, N.A., Young, B.E., Rodrigues, A.S.L., Fischman, D.L. and Waller, R.W., 2004. Status and trends of amphibian declines and extinction worldwide. *Science* **306**: 1783–86.

Svensk Naturförvaltning AB 2008. Uppföljning av stinkpaddans (Bufo calamita) populationsstatus längs Bohuskusten. (Evaluation of the population status of the Natterjack toad (*Bufo calamita*) along the Swedish west coast). Rapport 2008:41. Länsstyrelsen i Västra Götaland.

Tegelström, H. and Sjögren-Gulve, P., 2004. Genetic differentiation among northern European pool frog (*Rana lessonae*) populations. *Herpetological Journal* **14**: 187–93.

Voyles, J., Young, S., Berger, L., Campbell, C., Voyles, W.F., Dinudom, A., Cook, D., Webb, R., Alford, R.A. Skerratt, L.F. and Speare, R., 2009. Patogenesis of chytridiomycosis, a cause of catastrophic amphibian declines. *Science* **326**: 582–5.

Wirén, M., 2006. Grönfläckig padda I Sverige. Utvärdering av utförda artbevarandeåtgärder (1994–2005), förslag om framtida åtgärder samt artens tidigare and nuvarande förekomst. (The Green toad in Sweden: Evaluation of the conservation measures 1994–2005, suggestions on future measures and the earlier and present status). Rapport, Länsstyrelsen i Skåne län.

Wirén, M., 2010. Åtgärdsprogram för bevarande av grönfläckig padda 2011–2016. (Action plan for the Green toad 2011–2016). Rapport 6406, Naturvårdsverket.

63 Decline and conservation of amphibians in Finland

Ville Vuorio and Jarmo Saarikivi

Abbreviations and acronyms used in the text

BP	before present
CEF	continued ecological functionality
Ga	giga-annum, 1,000,000,000 years
GCN	great crested newt
m a.s.l.	metres above sea level

I. Introduction

Finland is situated in Northern Europe between the latitudes of 60° and 70° N. Landforms are defined by highly eroded Precambrian bedrock (age 3–1.5 Ga), while the present landscape is characterized by loose deposits from the last deglaciation stage some 12,000–9,000 years BP. The highest point in Finland is 1,328 m a.s.l., but about 80% of the area of Finland is low-lying land below 200 m a.s.l. (Tikkanen 2002).

The climate is either maritime or continental, depending on the direction of the air flow. The mean temperature varies notably from south to north. During summer, the mean temperature is 15°–12°C, respectively and during the winter 5°–13°C below zero, respectively. Lakes freeze normally in late November and become ice-free in April–May or June in the north. Rainfall is moderate during all seasons, 600–700 mm being the annual average (Finnish Meteorological Institute 2015).

Most of Finland (338,000 km²) belongs to the boreal zone; only the southwestern corner of the country is hemiboreal (Ahti *et al.* 1968). Forests cover 73%, water and wetlands 16%, and croplands and grasslands 9% of the territory (Eurostat 2015).

A. Amphibian species in Finland

Finland has five native amphibian species and seven species found more recently that are possibly of alien origin (Table 63.1). The common frog (*Rana temporaria*) is the most common amphibian species in Finland and it occurs throughout the whole country. The common toad (*Bufo bufo*) and the moor frog (*Rana arvalis*) are also common and widespread throughout the country, except for the northernmost area, where these two species occur sporadically and cannot be found in the fell areas north of the Arctic Circle. The smooth newt (*Lissotriton vulgaris*) is common in the southern part of the country, with the northernmost populations occurring in the Oulu region at the coast

Table 63.1   The status of amphibians in Finland. The placement of species in the EU Habitats Directive's different appendices has the following implications: II – conservation requires designation of special areas of conservation; IV – strictly protected species; V – limited means of capture, killing and transportation of individuals.

Species	Temporal Occurrence	Habitats Directive annex	The Red List of Finnish Species 2010
Triturus cristatus	native	II, IV	Endangered
Lissotriton vulgaris	native	–	Least concern
Bufo bufo	native	–	Least concern
Rana temporaria	native	V	Least concern
Rana arvalis	native	IV	Least concern
Ichthyosaura alpestris	2013 to the present	–	not evaluated
Pelophylax lessonae	2013 to the present	IV	Not evaluated
Pelophylax kl. *esculentus*	2013 to the present	V	Not evaluated
Pelophylax ridibundus	1937–1960 and 2008 to the present	V	Not evaluated
Hyla arborea	2015 to the present	IV	Not evaluated
Bombina variegata	2015 to the present	II, IV	Not evaluated
Bufotes viridis	1993–1994	IV	Not evaluated

of the Bay of Bothnia, bordering the northern part of the Baltic Sea. The great crested newt (GCN, *Triturus cristatus*) has a disrupted distribution occurring both in the Åland archipelago in the southwest and in Eastern Finland with a 400-km-wide gap in between. All the species are more abundant in the south of Finland and at least somewhat tolerant of brackish water (salinity < 1%). These five native species are protected in Finland.

Amphibian species and their abundance have been monitored by the Ministry of Environment for the Red List of Finnish Species, published so far four times: 1986, 1991, 2001 and 2010. The evaluations made by the expert groups have concluded that there is a declining trend in amphibian abundances. The most recent evaluation (Terhivuo and Mannerkoski 2010) is presented in Table 63.1.

II. Declining amphibians and species of special conservation concern

The only threatened native amphibian species in Finland is the GCN. It is classified as an endangered species (Terhivuo and Mannerkoski 2010) and its preferred habitat, forest ponds, is a threatened habitat (Ilmonen *et al.* 2008). The GCN occurs in Åland in the southwestern archipelago with *circa* 30 breeding ponds and in Eastern Finland with approximately 80 breeding ponds presently (Hertta 2015). Finnish threats to the species are due to the transfer of fish and the drying out and dredging of ponds (Terhivuo and Mannerkoski 2010).

Prior to 2004 there were 31 known ponds where the GCN bred in Eastern Finland. During 2004–2008, the North Karelian environmental centre (regional environmental authority) took part in a LIFE Nature project "Protection of *Triturus cristatus* in Eastern Baltic Region". As a result, there were 37 new breeding ponds found in Eastern Finland. During the project, the first artificial breeding ponds for the GCN were excavated (Briggs *et al.* 2006, Vuorio 2009). Between 2010 and 2013 there was a nationally funded GCN conservation project. During this project there were about 25 new ponds excavated and new conservation areas established around and between the existing breeding ponds. By the end of the breeding season in 2015, one third of the new artificial ponds were used for breeding by the GCN. The western population of the GCN in Åland (southwestern Finland) was surveyed in 2013 (Saarikivi 2013) and the species was found to be widespread in the archipelago with several known breeding ponds.

During recent years, knowledge of the ecology and habitat requirements of the GCN in the north has increased (Vuorio 2009; Vuorio *et al.* 2013, 2015). Clear felling in the vicinity of and between the breeding ponds causes barriers to dispersal, which have their greatest effects on small populations (Vuorio *et al.* manuscript under review in Annales Zoologici Fennici).

Climatic conditions play an important role in extirpations and colonizations in individual aquatic habitats (Hartel 2011). Climatic change is assessed to have considerable effects on the hydrology of ephemeral ponds (Brooks 2004, 2009; Leibowitz and Brooks 2008). The hydrology of the ponds in turn has been shown to have considerable effect on the hatching success of newts' eggs in the boreal zone (Vuorio *et al.* manuscript under review in Amphibia-Reptilia). The greatest present threats to populations of the GCN are posed by forestry activities and climatic change. If the snow cover is diminishing as estimated (Raisanen and Eklund 2012), there may in future be insufficient meltwater to maintain an adequate level of water in the breeding ponds.

III. Conservation measures and monitoring programmes

All amphibian species in Finland have been protected since 1983 by the Nature Conservation Decree. This protection means that it is forbidden to kill, disturb or harm any life stages of amphibians in any way. For example, the collecting of frogs' eggs and raising them to tadpoles is illegal in Finland. Alongside the national legislation are EU directives, namely the Habitats Directive, which classifies the species according to their conservation status.

The GCN is the only amphibian species in Finland that has a national action plan of its own (Vuorio 2009). The species and its habitats enjoy strict protection in Finland based on national (Nature Conservation Act, Water Act and Forest Act) and EU legislation (Habitats Directive).

The occurrence of amphibians in Finland was earlier charted by public surveys, which were organized by the Finnish Museum of Natural History (Terhivuo 1993). Since 2015 the collection of data is organized by the Museum's Amphibian and Reptile Atlas Project, which is still a public survey, but with developed web sites, on-line forms, identification guides, and active promoting and marketing. Since its launch in spring 2015, the Atlas Project has received some 2,200 amphibian observations in a season (half a year), which is far more than some dozens of observations collected yearly under the previous passive system. New data will be comparable with the earlier atlas prepared in the 1980s, 1990s and early 2000s and will provide an invaluable opportunity to study long-term changes in the distributions of individual species and for species diversity as a whole.

The risk of false identification is minimized by active administration and the observations can be confirmed by experts from photographs that often are attached to the database. Sensitive data on the occurrence of rare species or species of conservation concern are not available in the public version of the database.

IV. Non-native species

Since 2008, anuran species from the European green frog complex have been observed in the Turku region of southwestern mainland Finland. The marsh frog (*Pelophylax ridibundus*) was probably the first amphibian species to colonize urban coastal areas and ponds. The edible frog (*Pelophylax* kl. *esculentus*) and pool frog (*Pelophylax lessonae*) were observed and identified a few years later. These three species have since become established in the region, occurring currently in a number of ponds within 50 km from Turku (Hoogesteger *et al.* 2013). Also, the alpine newt (*Ichthyosaura alpestris*) was found in 2013 in a forest pond close to Turku. In 2015 even more new alien species were observed in the region. The European treefrog (*Hyla arborea*) was found in urban areas in Turku and Helsinki and the yellow-bellied toad (*Bombina variegata*) in the Turku region.

Earlier, occasional observations of non-native species had been very scarce. The marsh frog had introduced populations in Helsinki and Porvoo (southern Finland) in 1937–1960. One individual of the green toad (*Bufotes viridis*) was found in Kotka harbour (southeastern Finland) in 1993–1994.

The origin of these non-native species is unknown, but most likely they were deliberately introduced by humans. The rate at which new species have been found in recent years and the fact that observations are from the same region excludes natural pathways of migration. Furthermore, a new record of a reptilian species, the sand lizard (*Lacerta agilis*), has been found in the same region as well. Most of these recently observed new records around Turku are of species that are not invasive, not commonly kept as pets, and their natural ranges are 400–2,000 km from the newly inhabited sites. The intentional introduction of new species is prohibited in Finland and otherwise historically has involved mainly species of game animals.

V. Conclusions

Amphibians have adequate protection in Finland; the national conservation measures combine the protection of species with the protection of habitat. The situation is further improved with the Habitats Directive that has focused on species' habitats. Nevertheless, legislation does not improve the conditions for amphibians or considerably reduce loss of habitat, which is most likely the main reason for their decline. Changes in agricultural and silvicultural practices are needed in order to preserve wetlands and small bodies of water and their terrestrial surroundings – the habitat for amphibians. Often these aquatic habitats do not have to be in a pristine natural state in order to become valuable amphibian habitats. Small bodies of water such as ditches, private garden or golf-course ponds could be easily developed into amphibian sanctuaries by careful management – even in urban areas.

During the past decade there has been major improvement in conservation practices of the GCN. Every time there is a planned logging in the vicinity of a breeding pond, the logging area is defined together with the land owner, logging company and a conservation specialist. From 2013 onwards there is at least a 75-m buffer zone in which no forest practices are allowed.

The moor frog is both protected by national legislation and listed in the Habitats Directive. As the species is relatively common, conflicts of interests with various practitioners have resulted in several cases in which protection of moor frog habitat has prevented development. Nowadays, solutions in which resolution of conflicts between development and species protection are sought via a selection of species-conservation instruments. Measures of active conservation efforts such as CEF (continued ecological functionality) have shown their potential in some cases in which active conservation efforts are performed by digging new breeding ponds or modifying suitable habitats.

To ensure the survival of the GCN in the Finnish fauna it is necessary to guide forestry practices in the vicinity of the breeding ponds. There is still also a need to continue the excavation of new breeding ponds, especially in areas where the existing breeding ponds are prone to regular drying because of changing climatic patterns that change the hydrology of ponds. The exact selection of sites is of primary importance. Lately, GCN ponds have been dug according to three criteria: (1) there must be enough water in the pond throughout the larval period, (2) the pond should be deep enough to avoid complete freezing during winter months, and (3) the soil in the vicinity of the pond has to be rich enough, e.g. *Oxalis-Myrtillys* type or richer.

Most probably the new non-native species have been introduced by humans either by intentional release or accidentally via transportation in ships, containers or packages. The greatest risk of non-native species is the risk of their spreading diseases. So far Finland has remained free from *Rana*-virus, *Batrachochytrium dendrobatidis* and *B. salamandrivorans* (Riikka Holopainen / EVIRA, pers. comm. 24 August 2015).

VI. Acknowledgements

This study was funded by the Academy of Finland (decision 263465, Water management and peat production: from relevant facts to effective norms). Thanks are due to Dr. Markus Piha from the Museum of Natural History who provided valuable information about the amphibian atlas project. We also thank Dr. Olli-Pekka Tikkanen for his valuable comments on the manuscript.

VII. References

Ahti, T., Hämet-Ahti, L. and Jalas, J., 1968. Vegetation zones and their sections in northwestern Europe. *Annales Botanici Fennici* **5**: 169–211.

Briggs, L., Rannap, R., Pappel, P., Bibelriether, F. and Päivärinta, A., 2006. *Monitoring methods for the great crested newt* Triturus cristatus. Odense, Tallinn.

Brooks, R.T., 2004. Weather-related effects on woodland vernal pool hydrology and hydroperiod. *Wetlands* **24**: 104–14.

Brooks, R.T., 2009. Potential impacts of global climate change on the hydrology and ecology of ephemeral freshwater systems of the forests of the northeastern United States. *Climatic Change* **95**: 469–83.

Eurostat. 2015. Statistics Explained. http://ec.europa.eu/eurostat/statistics-explained/index.php/Main_Page. Accessed 24 August 2015.

Finnish Meteorological Institute. 2015. Vuositilastot. http://ilmatieteenlaitos.fi/vuositilastot. Accessed 24 August 2015.

Hartel, T., Bancila, R. and Cogălniceanu, D., 2011. Spatial and temporal variability of aquatic habitat use by amphibians in a hydrologically modified landscape. *Freshwater Biology* **56**: 2288–98.

Hertta. 2015. Species application of the environmental data system. Finnish Environment Institute.

Hoogesteger, T., Rahkonen, J. and Karhilahti, A., 2013. Pool frog (*Pelophylax lessonae*) Camerano 1882 (Anura, Ranidae), an addition to the Finnish amphibian fauna. *Memoranda Societatis pro Fauna et Flora Fennica* **89**: 25–31.

Ilmonen, J., Leka, J., Kokko, A., Lammi, A., Lampolahti, J., Muotka, T., Rintanen, T., Sojakka, P., Teppo, A., Toivonen, H., Urho, L., Vuori, K.M. and Vuoristo, H., 2008. Sisävedet ja rannat. In A. Raunio, A. Schulman and T. Kontula, eds. *Suomen luontoyyppien uhanalaisuus – osa 1*. Suomen ympäristökeskus, Helsinki.

Leibowitz, S.G. and Brooks, R.T., 2008. Hydrology and landscape connectivity of vernal pools. In A.J.K. Calhoun and P.G. DeMaynadier, eds. *Science and Conservation of Vernal Pools in Northeastern North America*. CRC Press, Boca Raton.

Raisanen, J., and Eklund, J., 2012. 21st Century changes in snow climate in Northern Europe: a high-resolution view from ENSEMBLES regional climate models. *Climate Dynamics* **38**: 2575–91.

Saarikivi, J. 2013. Ahvenanmaan rupiliskoselvitys vuonna 2013. Faunatica Oy.

Terhivuo, J. 1993. Provisional atlas and status of populations for the herpetofauna of Finland in 1980–92. *Annales Zoologici Fennici* **30**: 55–69.

Terhivuo, J., and Mannerkoski, I., 2010. Matelijat ja sammakkoeläimet. In P. Rassi, E. Hyvärinen, A. Juslén and I. Mannerkoski, eds. *The 2010 Red List of Finnish Species*. Ympäristöministeriö & Suomen ympäristökeskus, Helsinki.

Tikkanen, M., 2002. The changing landforms of Finland. *Fennia* **180**: 21–30.

Vuorio, V., 2009. Suomen uhanalaisia lajeja: Rupilisko (*Triturus cristatus*). Pohjois-Karjalan ympäristökeskus, Joensuu.

Vuorio, V., Heikkinen, R.K. and Tikkanen, O.P., 2013. Breeding success of the threatened great crested newt in boreal forest ponds. *Annales Zoologici Fennici* **50**: 158–69.

Vuorio, V., Tikkanen, O.P., Mehtatalo, L. and Kouki., J., 2015. The effects of forest management on terrestrial habitats of a rare and a common newt species. *European Journal of Forest Research* **134**: 377–88.

64 Decline and conservation of amphibians in Estonia

Riinu Rannap

Abbreviations and acronyms used in the text

UTM   *Area between longitudes 24° and 30°E and longitudes 0° and 84°N*

I. Introduction

Estonia (total area 45,227 km²) is located in Northern Europe, on the eastern coast of the Baltic Sea. The country is situated in the northern part of the temperate zone and in the transition between maritime and continental climates. The climate of coastal and inland areas varies significantly due to the influence of the Baltic Sea. Lying on the level northwestern part of the East European Platform, Estonia is relatively flat and low with the average elevation reaching only 50 meters. Uplands (usually 75–100 m above sea level) and plateau-like areas alternate with lowlands, depressions, and valleys (Statistics Estonia 2017).

Annual precipitation in Estonia exceeds evaporation, allowing, with help from the relief of the region, the formation of numerous inland bodies of water. Lakes account for 4.7% of the Estonian territory and are unevenly distributed over the country. Most of the lakes are shallow with only a few (approximately 20) of them being deeper than 20 m (Statistics Estonia 2017).

Estonia belongs to the northern part of the mixed forest sub-zone of the temperate forest zone. Owing to a large variety of soil conditions and moisture regimes, many different types of forests grow there, collectively covering more than half of the Estonian territory. Approximately one

Table 64.1   The species of amphibians present in Estonia and their status

Species	EU Habitats Directive	Protection category in Estonia	Estonian Red Data Book
Lissotriton vulgaris	–	III	Least concern
Triturus cristatus	II, IV	II	Vulnerable
Pelobates fuscus	IV	II	Vulnerable
Bufo bufo	–	III	Least concern
Bufo calamita	IV	I	Endangered
Bufo viridis	IV	I	Critically endangered
Rana temporaria	V	III	Least concern
Rana arvalis	IV	III	Least concern
Pelophylax kl. esculentus	–	III	Least concern
Pelophylax lessonae	IV	III	Least concern
Pelophylax ridibundus	–	III	Data deficient

Fig. 64.1 *Bufo calamita* (A) during the breeding season at one of the last coastal meadows on Manilaid Island, western Estonia (B). Photographs by Riinu Rannap.

fourth of the territory is occupied by peatlands, the most widespread type being raised bog, in which a thick layer of peat has accumulated over millennia (Statistics Estonia 2017).

As a result of human activity, abundant meadows have developed in Estonia. Meadows are classified into several types according to their soil and water regimes. Dry and fresh meadows, which have developed on the sites of former boreo-nemoral forests, are relatively rich in species. Alvar meadows, occurring on thin soils on limestone, however, have become very rare owing to a decline in traditional agricultural practices (Helm *et al.* 2006). The same is true for floodplain meadows, coastal meadows and wooded meadows (sparse semi-natural woodlands formed as a result of regular mowing and grazing) (Luhamaa *et al.* 2001; Truus and Tõnisson 2009; Kana *et al.* 2015).

Fig. 64.2 *Pelobates fuscus* (Piirassaar Island, Peipus Lake, eastern Estonia) (A) and two of its other natural breeding sites: (B) a beaver dam in Karula National Park, southern Estonia, and (C) Karst Lake, a temporary body of water in northern Estonia. *Rana arvalis*, *Rana temporaria*, *Lissotriton vulgaris* and *Pelophylax lessonae* also breed at B and *Rana arvalis*, *Rana temporaria*, *Lissotriton vulgaris* and *Triturus cristatus* also breed at C. Photograph A by Merike Linnamägi, B by Riinu Rannap, and C by Maris Markus.

Amphibians are represented by 11 species in Estonia (Table 64.1). Three species – the green toad (*Bufo [Pseudepidalea] viridis*), the natterjack toad (*B.[Epidalea] calamita* Figure 64.1) and the common spadefoot toad (*Pelobates fuscus* Figure 64.2) have the northern edge of their range within the country. The most widespread and common amphibians are the smooth newt (*Lissotriton vulgaris*), the common toad (*Bufo bufo*), the common frog (*Rana temporaria*) and the moor frog (*R. arvalis*). During past decades the pool frog (*Pelophylax lessonae*) also has remarkably expanded its range in Estonia, spreading from the southern part of the country towards the north. All amphibian species are under protection in Estonia (Table 64.1).

II. Declining species of amphibians and species of special conservation concern

Four amphibian species – the green toad, the natterjack toad, the common spadefoot toad and the northern crested newt – are declining in Estonia.

The green toad used to inhabit the eastern and southeastern parts of the country, particularly the coastal plains of Lake Peipus and Piirisaar Island. It occurred in coastal villages and on flood plain meadows, the latter being used as pastures for cattle grazing or for making hay. During the twentieth century the green toad has declined substantially, mainly owing to the destruction of its habitats. The coastal flood plains – the main breeding grounds for the toad – have become overgrown with tall vegetation (willows and reed), owing to lack of sufficient land management (grazing, mowing). At present, there are no viable populations of the green toad known in Estonia.

The natterjack toad occurs mainly on the western coast and on the western archipelago of Estonia, where the climate is notably milder than in the central parts of the country. The toad inhabits open sand dunes and coastal meadows, used traditionally as pastures (Figure 64.1). In the first half of the 20th century, the natterjack toad was rather common in those areas. Between the 1930s and the 2000s, however, 73% of the local populations have been lost. At the same time, the number of localities with coastal meadows has declined by 91% (Rannap *et al.* 2007). Currently only 15 isolated populations of the natterjack toad remain in Estonia (Figure 64.3). The main reason behind this decline is the destruction of habitat – due to large-scale adverse impacts of intensified agriculture, drainage (Beintema 1991) and lack of management (Koivula and Rönkä 1998; Kuresoo and Mägi 2004; Ottvall and Smith 2006); the Baltic coastal meadows are now amongst the most threatened habitats in Europe. Since the 1960s, coastal eutrophication has led to rapid expansion of reeds on coastal meadows (Soikkeli and Salo 1979) and extensive drainage has dried the meadows, destroying their heterogeneity and leading to the disappearance of natural depressions and shallow ponds. The most drastic changes, however, appeared as a result of the land reform in 1990, when collective farms collapsed and new landowners, for economic reasons, often ceased to use coastal meadows. As coastal pastures and hayfields of low productivity had already been gradually excluded from use since the 1950s, the reform resulted in most coastal meadows being cast aside, quickly overgrowing with high vegetation and scrub. Between 1980 and 2000, the number of cattle declined three-fold and the number of sheep five-fold (Luhamaa *et al.* 2001); the overall area of managed coastal meadows has decreased from 29,000 ha to 8,000 ha over the past 50 years (Luhamaa *et al.* 2001). At the same time, open dune areas have largely been forested and are now covered with pine plantations. In addition, habitat destruction – overgrowing and afforestation – has caused increasing predation by reducing and fragmenting open areas and creating suitable hiding places and den-sites for foxes (*Vulpes vulpes*) and racoon dogs (*Nyctereutes procyonoides*). In Estonia, the latter is an invasive alien species whose population has increased tremendously during the past few decades. Such increase in numbers has brought along extensive colonization of coastal areas, islands and even small islets by racoon dogs, causing serious decline

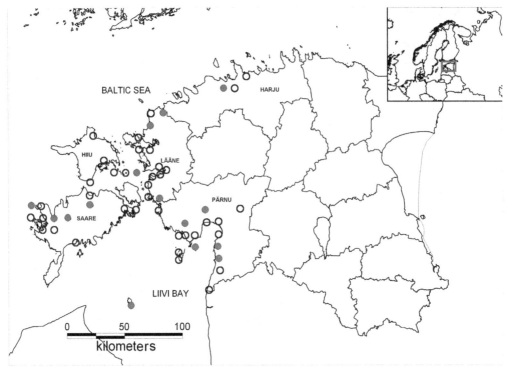

Fig. 64.3 Distribution of the natterjack toad (*Bufo calamita*) in the 1930s (hollow symbols) and 2000s (filled symbols) in Estonia.

and even extirpation of amphibians. This was the case on Kumari islet in Matsalu National Park where, within two years, racoon dogs eradicated entire populations of the natterjack toad, the common frog and the moor frog.

The northern crested newt and the common spadefoot toad have also witnessed a severe decline in Estonia during past centuries. Historically, both species were widely distributed and numerous in southern and southeastern Estonia, where the landscape is hilly with mosaics of forests, grasslands, extensively used farmlands and a great number of different bodies of water: small lakes, beaver dams, natural depressions, cattle-watering ponds, garden ponds, sauna ponds and ponds historically used for soaking flax. During the second half of the twentieth century, however, cattle farming decreased, resulting in many small ponds overgrowing with scrub. Additionally, several bodies of water were filled in or fish were introduced (Rannap *et al.* 2009).

In addition to inhabiting hilly southern Estonia, both species also occur in the karst areas (Figure 64.2) in the central and northern parts of the country. In these regions the limestone is close to the surface and therefore topographically higher than in the surrounding area, causing intense filtration and karst processes. During the spring, melting snow and rainfall fill several depressions and cavities with water, forming temporary lakes. Slow filtration of the water follows, resulting in the depressions becoming dry by late summer/early autumn. These types of large but temporary bodies of water are excellent breeding habitats for both species. Owing to intensification of agriculture, however, many of these temporary lakes have been filled in or destroyed by amelioration. Moreover, because of intensive agriculture, many populations of the northern crested newt and the common spadefoot toad have become small and isolated in this area.

III. Conservation measures and monitoring programmes

Large-scale conservation of amphibians in Estonia began in 2000. The first target species was the natterjack toad, which then had only 15 small isolated populations left. The main conservation goal was to secure the existing populations and increase the number of individuals in each population through management and restoration of habitat.

To achieve this goal, several national and international projects were launched, among these the LIFE-Nature project "Boreal Baltic Coastal Meadow preservation in Estonia", which targeted the restoration and maintenance of the natterjack toad's coastal meadow habitats. Complying with the project, grazing was initiated or intensified on several coastal meadows – current and former natterjack toad's habitats. This was achieved by purchasing cattle and sheep for local farmers. Also, several work camps were organized to clear reeds and bushes from coastal areas. Importantly, in 2001 the Estonian Ministry of Environment started to pay subsidies to local farmers and land owners to restore and maintain semi-natural habitats, including coastal meadows. In sand-dune areas, pine plantations were cut down and shallow ponds for breeding created to restore the natterjack toad's preferred open, sunny habitats. In addition to habitat management and restoration, a supportive breeding program was carried out in 12 populations of the toad to secure annual breeding success. Since 2004 the protection and habitat management of the natterjack toad has been implemented according to the species' National Action Plan approved by the Minister of Environment. During the past decade conservation actions have been carried out in 13 natterjack toad localities. At present, 11 (73%) populations are increasing or stable, but four are still in decline.

Actions towards conserving the green toad were carried out in 2001 when three former breeding ponds were restored (cleaned of scrub and mud) on Piirissaar Island where the last viable population of the species was found. Unfortunately, no colonization of those ponds by the green toads was observed, indicating that the conservation management for this species had come far too late.

To halt the decline of the northern crested newt and the common spadefoot toad, save their small and isolated populations from extinction, and restore several viable meta-populations of those species, two international LIFE-Nature projects "Protection of *Triturus cristatus* in the Eastern Baltic Region" and "Securing *Leucorrhinia pectoralis* and *Pelobates fuscus* in the northern distribution area in Estonia and Denmark" were launched in 2004 and 2009 respectively. As the main reason behind the decline of the northern crested newt and the common spadefoot toad is a lack of high-quality breeding sites (Rannap *et al.* 2009), our main goal was to restore and create such habitats. Since then 280 ponds have been restored or created in southern and northern Estonia in the distribution area of the target species. In order to increase the probabilities of colonization and preserve existing populations, ponds were constructed in clusters (at least three ponds in each), with distances between them being no more than 500 m and at least one constructed pond being located within 200 m of a source pond of the target species. By 2008, successful breeding of the northern crested newt had been recorded in 23 of 25 clusters (92%), and of the common spadefoot toad in 17 of 21 clusters (81%) (Rannap *et al.* 2009). In addition to the target species, all other amphibians present in these regions also benefited from the restoration and construction of ponds. The National Action Plan for the northern crested newt was launched and approved by the Minister of Environment in 2008 and for the common spadefoot toad in 2015. Since then the conservation actions and national projects have been arranged by the Estonian Environmental Board. National Action Plans for the green toad is in process.

All Estonian amphibian species are under a state-monitoring program as of 1997. A separate program has been launched for the natterjack toad, covering all of its populations. The rest of the species are monitored annually within 20 UTM squares (including 100 small bodies of water) all over Estonia. The main methods of monitoring are:

- counting of calling males and strings of eggs of the natterjack toad;
- searching for eggs and dip-netting of larvae – to determine not only the presence of species but also their breeding success.

IV. Conclusions

Four of the 11 Estonian amphibian species – the green toad, the natterjack toad, the common spadefoot toad and the northern crested newt are in decline. Importantly, three of those species occur at the northern edges of their ranges in Estonia. The main reason behind their decline is destruction of habitat. The natterjack toad and the green toad lack open, sunny habitats on flood-plain meadows, coastal meadows or sand dunes (Rannap et al. 2007). The common spadefoot toad and the northern crested newt mostly lack high-quality breeding sites exposed to the sun (Rannap et al. 2012, 2013). Large-scale, landscape-level conservation work, carried out since 2000 in Estonia has secured many populations of these threatened species; this work, however, needs to be continued.

V. References

Beintema, A.J., 1991. What makes a meadow bird a meadow bird? *Wader Study Group Bulletin* **61**: Supplement 3–5.

Helm, A., Hanski, I. and Pärtel, M., 2006. Slow response of plant species richness to habitat loss and fragmentation. *Ecology Letters* **9**: 72–7.

Kana, S., Otsus, M., Sammul, M., Laanisto, L. and Kull, T., 2015. Change in species composition during 55 years: a re-sampling study of species-rich meadows in Estonia. *Annales Botanici Fennici* **52**: 419–31.

Koivula, K. and Rönkä, A., 1998. Habitat deterioration and efficiency of antipredator strategy in a meadow-breeding wader, Temminck's stint (*Calidris temminckii*). *Oecologia* **116**: 348–55.

Kuresoo, A. and Mägi, E., 2004. Changes of bird communities in relation to management of coastal meadows in Estonia. In *Coastal Meadow Management*, eds R. Rannap, L. Briggs, K. Lotman, I. Lepik and V. Rannap. Prisma Print, Tallinn.

Luhamaa H., Ikonen I. and Kukk T., 2001. *Seminatural communities of Läänemaa County, Estonia*. Society of Protection of Seminatural Communities, Tartu – Turku.

Ottvall, R. and Smith, H.G., 2006. Effects of an agri-environment scheme on wader populations of coastal meadows of southern Sweden. *Agriculture, Ecosystems and Environment* **113**: 264–71.

Rannap R., Lõhmus, A. and Briggs, L., 2009. Restoring ponds for amphibians: A success story. *Hydrobiologia.* **634**: 87–95.

Rannap, R., Lõhmus, A. and Jakobson, K., 2007. Consequences of coastal meadow degradation: the case of the natterjack toad (*Bufo calamita*) in Estonia. *Wetlands* **27**: 390–8.

Rannap, R., Lõhmus, A. and Linnamägi, M., 2012. Geographic variation in habitat requirements of two coexisting newt species in Europe. *Acta Zoologica Academiae Scientiarum Hungaricae* **58**: 73–90.

Rannap, R., Markus, M. and Kaart, T., 2013. Habitat use of the Common Spadefoot Toad (*Pelobates fuscus*) in Estonia. *Amphibia-Reptilia* **34**: 51–62.

Soikkeli, M. and Salo, J., 1979. The bird fauna of abandoned shore pastures. *Ornis Fennica* **56**: 124–32.

Statistics Estonia 2017 (http://www.stat.ee/statistics).

Truus, L. and Tõnisson, A., 1998. The ecology of floodplain grasslands in Estonia. In *European wet grasslands; biodiversity, management and restoration*, eds. C.B. Joice and P.M. Wade. Wiley, New York.

65 Decline and conservation of amphibians in Latvia

Aija Pupina, Mihails Pupins, Andris Ceirans and Agnese Pupina

Abbreviations and acronyms used in the text and references

Bd	Batrachochytrium dendrobatidis
Bsal	Batrachochytrium salamandrivorans
IUCN	*International Union for the Conservation of Nature*
GE	*genomic equivalents*
RDB	*Red Data Book of Latvia*
HD	*Habitats Directive*
LEGM	*Latvian Environment, Geology and Meteorology Centre*
MKN	*Ministru Kabineta Noteikumi (the documents, approved by the Cabinet of Ministers of Latvia)*
qPCR	*quantitative polymerase chain reaction or real-time polymerase chain reaction*

I. Introduction: The country and its amphibian fauna

Latvia – officially the Republic of Latvia (in Latvian: Latvija, Latvijas Republika) – is a country of the Baltic region of northern Europe and lies on the east coast of the Baltic Sea and its Gulf of Riga. Latvia has been a member of the European Union since 2004. The total area of Latvia is 64,589 km²; its length from north to south is 210 km and from east to west is 450 km. Latvia has land borders with Estonia in the north (343 km), with Lithuania in the south (588 km), with Belarus in the southeast (161 km), and with Russia in the east (276 km) (Central Statistical Bureau Republic of Latvia 2012). Latvia also shares its borders with Estonia, Sweden and Lithuania in the Baltic Sea. A relatively large proportion of Latvia's perimeter, relative to its surface area, transects various natural habitats also shared by neighbouring countries. This ecological continuity with adjacent countries is reflected in similarities in their amphibian faunas.

The largest part of the territory of Latvia is at a height of not more than 100 m above sea level. Latvia has a temperate seasonal climate that can also be described as oceanic/maritime (Central Statistical Bureau Republic of Latvia 2012). Latvia has four well-defined seasons of approximately equal duration. In winter the average temperature is –6°C, with minima of –20° to –30°C, usually lasting for several weeks; winter conditions include a short day-length, a constant mantle of snow up to 0.7 metres deep, total freezing of small ponds, and persistent ice 15–20 cm thick on larger ones. Such conditions affect survival of amphibians overwintering either underwater or on land, especially in their first year of life, and declines occur during winter. The average air temperature is 19°C in summer, with maxima of 25°–30°C commonly occurring, especially in the country's interior. It often rains, especially in the second part of the summer (Latvian Biodiversity Clearing-House Mehanism 2015) (Figure 65.1), thereby creating numerous large and small ponds and other wetland habitats used by amphibians.

There is a moderate marine climate in the western coastal regions of Latvia, with warmer winters and cooler summers than in the eastern continental part of the country; this regional difference in climate influences the distribution of the amphibian fauna within Latvia.

Of the total surface of Latvia, agricultural land covers 23, 909 km² (44.0%) and forest 29, 996 km² (55.2%); inland fresh waters cover the remaining 419 km² (0.8%). The largest lake is Lubans with a surface area of 80.7 km²; the deepest lake is Dridzis with a depth of 65.1 m. The longest river in Latvia is the Gauja, which is 452 km long. The longest river flowing through the territory of Latvia from Belarus is the Daugava, which has a total length of 1,005 km of which 352 km are within the territory of Latvia. There are many lakes, rivers, large and small ponds and other small bodies of water used by amphibians as habitats and migratory routes.

The average density of the human population is 34.3/km². There are five economic/political planning regions in Latvia: Kurzeme, Latgale, Riga, Vidzeme and Zemgale. They differ in relief, climate and level of economic and agricultural development, as well as in urban saturation which is important for the composition and distribution of the amphibian fauna.

Nowadays amphibians are represented by 13 permanently resident species in Latvia – 11 anurans and two caudates; all have Least Concern IUCN Red List Status for Europe, but with some species exhibiting changes in population density in Latvia (Table 65.1).

Latvia is the most northern country of the European Union in which the European fire-bellied Toad *Bombina bombina* resides, and the species reaches the northern extreme of its distribution there (Kuzmin *et al.* 2008). Latvia is also the most northern extent in the EU of the European treefrog (*Hyla arborea*). *Hyla arborea* was officially declared as extirpated in Latvia but was successfully reintroduced into the country using zoocultures established from animals caught in Belarus (Zvirgzds 2003).

Four species – the green toad *Bufotes viridis*, the natterjack toad *Epidalea calamita*, the common spadefoot toad *Pelobates fuscus* and the northern crested newt *Triturus cristatus* – reside in Latvia near the northern borders of their European distribution. *Epidalea calamita* is encountered mainly near the coast of the Baltic Sea and in central Latvia; over the past few years it has been observed in the southeastern part of the country as well. The other three of these species are encountered

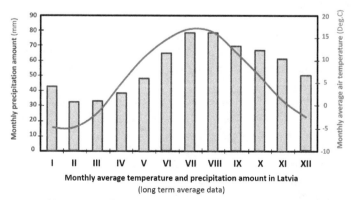

Fig. 65.1 Long-term monthly averages of air temperature and precipitation in Latvia (Latvian Environment, Geology and Meteorology Centre 2015).

Table 65.1 The species of amphibians present in Latvia, their IUCN status for Europe (IUCN 2014) and the trends of their populations in Latvia.

Species	English Name	Latvian Name	IUCN Red List Status (Europe)	IUCN Population trend (Europe)	Population trend in Latvia
Lissotriton vulgaris (Linnaeus, 1758)	Smooth Newt	Parastais tritons	Least Concern	stable	stable
Triturus cristatus (Laurenti, 1768)	Northern Crested Newt	Lielais tritons	Least Concern	decreasing	decreasing
Bombina bombina (Linnaeus, 1761)	European Fire-bellied Toad	Sarkanvēdera ugunskrupis	Least Concern	decreasing	decreasing generally, but increasing at some localities
Hyla arborea (Linnaeus, 1758)	European Tree Frog	Parastā kokvarde	Least Concern	decreasing	increasing
Pelobates fuscus (Laurenti, 1768)	Common Spadefoot Toad	Brūnais varžukrupis	Least Concern	decreasing	decreasing
Bufotes viridis (Laurenti, 1768)	European Green Toad	Zaļais krupis	Least Concern	decreasing	unknown
Epidalea calamita (Laurenti, 1768)	Natterjack Toad	Smilšu krupis	Least Concern	decreasing	decreasing
Bufo bufo (Linnaeus, 1758)	Common Toad	Parastais krupis	Least Concern	stable	stable
Pelophylax lessonae (Camerano, 1882)	Pool Frog	Dīķa varde	Least Concern	decreasing	stable
Pelophylax kl. esculentus (Linnaeus, 1758)	Edible Frog	Zaļa varde	Least Concern	decreasing	stable
Pelophylax ridibundus (Pallas, 1771)	Eurasian Marsh Frog	Ezera varde	Least Concern	increasing	unknown
Rana arvalis (Nilsson, 1842)	Moor Frog	Purva varde	Least Concern	stable	stable
Rana temporaria (Linnaeus, 1758)	Common Frog	Parastā varde	Least Concern	stable	stable

throughout most of Latvia, except in the north, but their populations are mostly small and fragmented.

Four other Latvian amphibians are abundant over the entire country and occur in a wide range of habitats: the smooth newt *Lissotriton vulgaris*, the common toad *Bufo bufo*, the common frog *Rana temporaria* and the moor frog *Rana arvalis*. *Pelophylax* spp. are also widely abundant throughout Latvia and are even encountered in urbanized habitats (Rimicans *et al.* 2010; Pupins and Pupina 2011a) (Table 65.2).

II. Threats to amphibians in Latvia

According to the official national requirements for development of Species Conservation Plans, factors posing threats to the habitats, ecosystems and populations of the protected species in Latvia can be divided into those of natural origin and those of anthropogenic origin. These can, of course, act synergistically. Threats of either kind are of less concern for amphibians that are widespread in Latvia, however, than they are for species that are represented by small or fragmented populations.

A relatively cold climate is a permanent natural cause of paucity and limited abundance of amphibian species in Latvia, such as *Triturus cristatus*, *Bombina bombina*, *Hyla arborea*, *Bufotes viridis*, *Epidalea calamita* and *Pelobates fuscus*. Of course, all species of Latvian amphibians are generally adapted to the Latvian climate as confirmed by their persistence there over a long period of time. However, the Latvian climate does pose challenges to amphibians in terms of achieving thermo-optimality and finding suitable habitats for feeding, breeding (Pupina and Pupins 2009) and overwintering. Therefore, the cold (−20° to −30°C) Latvian winters with relatively thin layers of snow are dangerous for amphibians overwintering in terrestrial sites, such as burrows or among

Table 65.2 The species of amphibians present in Latvia and their distribution and status in the country (Caune 1992; Pupins and Pupina 2011a; Pupina and Pupins 2012).

Species	Status in Latvia	Distribution in Latvia
Lissotriton vulgaris	common	whole territory
Triturus cristatus	rare	whole territory
Bombina bombina	very rare	southern part of Latvia
Hyla arborea	very rare	western part of Latvia (reintroduced)
Pelobates fuscus	rare	south and central territory, excluding the north of Latvia
Bufotes viridis	rare	south and central territory, excluding the north of Latvia
Epidalea calamita	very rare	central and western part of Latvia, near the Baltic sea; southeastern part of Latvia
Bufo bufo	common	whole territory
Pelophylax lessonae	common	whole territory
Pelophylax kl. esculentus	common	whole territory
Pelophylax ridibundus	common	south and central territory, excluding the north of Latvia
Rana arvalis	common	whole territory
Rana temporaria	common	whole territory

roots, and may lead to high mortality. Additionally, during such winters small ponds freeze all the way to the bottom and, in deeper ones, the oxygen regimen worsens, thereby causing the death of hundreds of amphibians overwintering there. Cool, short summers often occur in Latvia, with the result that some tadpoles, especially of southern species like *Bombina bombina*, do not have time to go through metamorphosis and gain sufficient weight before overwintering (A. Pupina and M. Pupins, unpublished data). In the relatively cool Latvian climate, amphibians mostly choose small, stagnant ponds, well heated by the sun, for reproduction. Therefore, when a dry, hot summer occurs, such ponds often dry up, resulting in mass mortality of amphibian larvae (A. Pupina, M. Pupins, unpublished data) (Figure 65.2).

Long-term anthropogenic transformation of habitats for agricultural use led to a reduction in the number of ponds in Latvia, and in impairment of habitat suitable for amphibians. The dams and ponds of beavers (*Castor fiber*), which are optimal for amphibians in Latvia, also are often destroyed by land-owners (Pupins and Pupina 2007). Drainage of wetlands, deforestation and formation of dams along riversides, straightening of small rivers and their subsequent transformation into soil-reclamation canals have been traditionally carried out over several decades in Latvia. These works were especially developed during the Soviet occupation of Latvia and led to the disappearance of tens of thousands of bodies of water, including small ponds, especially in the central part of Latvia. Phreatic decline, caused by melioration, and the wide use of fertilizers in agriculture during the Soviet time led to rapid eutrophication and to many of the remaining ponds becoming choked with weeds, which as a result became too dark and cold, and thus inappropriate for amphibians (Pupins and Pupina 2006; Pupina *et al.* 2012). Also, aquatic and terrestrial habitats continue to be overgrown by bushes and canes nowadays because of deagrarianization, which is becoming the main reason for degradation of amphibian habitats in Latvia.

Mortality on roads is an important negative factor for Latvian amphibians, especially during the spring and autumnal mass migrations, and for small populations of declining species. For example, in spring 2014 the authors counted 224 flattened adult and young *Rana temporaria* in Daugavpils on a 200-m stretch of asphalt road between breeding and wintering habitats. Also, 26 flattened or injured adult *Bufo bufo*, most of them in amplexus (Figure 65.3), were found on a 50-m section of unsealed road near the town of Akniste. There are no traffic signs in Latvia warning drivers of amphibian crossings, and barrier fences guiding amphibians to tunnels under the roads

Fig. 65.2 Drying *Pelophylax* sp. tadpoles in a pond where *Pelobates fuscus*, *Triturus cristatus*, and *Bombina bombina* are also living (Katriniski, Demene, Daugavpils novads; 55° 43' 10.56" N; 26° 32' 54.51" E). Photograph by Aija Pupina.

Fig. 65.3 Mortality on a road: flattened and injured adult *Bufo bufo* in amplexus (Akniste; 56°10' 8.61" N; 25° 43' 16.25" E). Photograph by Mihails Pupins.

Fig. 65.4 Burning old grass on the shore of a pond used by *Bombina bombina*, *Pelobates fuscus*, *Lyssotriton vulgaris* and *Pelophylax* sp. (Daugavpils novads; 55° 50′ 5.62″ N; 26° 29′ 3.43″ E). Photograph by Agnese Pupina.

are not used; therefore, mortality on roads is very high and can lead to fragmentation or extirpation of small populations.

Burning of dried grass is widely practised in Latvia. Such burning is carried out in the early spring on warm, sunny days and is very dangerous for amphibians that are migrating from overwintering sites to ponds for reproduction (Pupins and Pupina 2006). The authors have many times observed such fires near the edges of ponds inhabited by the declining species *Triturus cristatus*, *Bombina bombina*, *Pelobates fuscus* and some others (Figure 65.4). The authors have a dozen times found live amphibians with half-burnt bodies, injured by the burning of grass (Figure 65.5).

The degradation and disappearance of the majority of factories and agricultural firms in Latvia after the end of the Soviet occupation led to the appearance of a great number of derelict buildings, communication lines and shafts acting as mortal traps for amphibians. We managed to count 38 atrophied individuals of amphibians in just one shaft in Kumbuli, Daugavpils region.

There are a great number of illegal dumps around many Latvian villages, i.e. in forests, slashes, drains, marshes and bodies of water, which are the main habitats of amphibians. These dumps are the source of pollution of the environment of amphibians (Figure 65.6) wherein they are also injured by cuts from broken glass (Figure 65.7).

Fig. 65.5   Effect of burning grass: adult *Rana temporaria* burned while migrating to a breeding pond (Daugavpils novads; 55° 50′6.11″ N; 26° 29′4.60″ E). Photograph by Agnese Pupina

Fig. 65.6   Rubbish (empty bottle) and *Bombina bombina* in the breeding biotope of *Rana temporaria*, *Triturus cristatus*, *Pelobates fuscus*, *Pelophylax* sp., and *Bombina bombina* (Katriniski, Demene, Daugavpils novads; 55° 43′ 10.56″ N; 26° 32′54.51″ E). Photograph by Agnese Pupina.

Fig. 65.7 Adult *Rana temporaria* showing wound caused by broken glass in a rubbish dump (Kalkunes pagasts, Daugavpils novads; 55° 50′ 14.04″ N; 26° 29′ 27.25″ E). Photograph by Mihails Pupins.

Fig. 65.8 This invasive fish *Perccottus glenii* found in a breeding pond of six amphibian species (Microreserve Ilgas, territory Natura 2000, Skrudaliena pagasts, Daugavpils novads; 55° 41′ 32.88″ N; 26° 46′ 14.61″ E). Photograph by Mihails Pupins.

Invasive predators are dangerous for amphibians in Latvia, one of which is a fish, the Chinese sleeper (*Perccottus glenii*) (Figure 65.8). Having become established in small, warm bodies of water, which are the main habitats used by amphibians for reproduction in Latvia, this species can fully destroy or seriously damage the larvae of amphibians (Pupins and Pupina 2012a). *Perccottus glenii* was found by the authors in catchment basins of almost every large river and in many lakes of Latvia (Pupins and Pupina 2012b); it was found in the territories of many populations of *Bombina bombina* and *Triturus cristatus*, and continues to extend its distribution (Pupina and Pupins 2014; Pupina *et al.* 2015).

Another invasive predator in Latvia regarded as dangerous for larvae of amphibians is the widely spread fish *Carassius auratus*, whose natural range is eastern Asia. This species was imported into Latvia in 1948 and released before 1990 into at least 180 (23%) of Latvian lakes and into many other bodies of water (Andrusaitis 2003). We have discovered that in Latvia amphibians avoid using bodies of water for reproduction if fish are present (personal observations). Introduction of invasive and local fish into water inhabited by amphibians is carried out by local citizens and by anglers; *Perccottus glenii* is sold on the Internet (A. Pupina, M. Pupins, and Ag. Pupina, unpublished data).

There are seven species and subspecies of exotic freshwater turtles that we recorded in the wild in Latvia: *Trachemys scripta troostii, Mauremys caspica, Mauremys rivulata, Pelodiscus maackii, Trachemys scripta elegans* (Pupins and Pupina 2011b), *Trachemys scripta scripta, Graptemys pseudogeografica kohnii* (M. Pupins, A. Pupina, unpublished data); mainly single individuals of *Trachemys scripta elegans* successfully overwinter and attempt oviposition (Pupins 2007; Pupins and Pupina 2012c). These carnivorous species can be dangerous for small populations of declining species of amphibians, especially in the south of the country (Pupins and Pupina 2006), and they can also introduce exotic parasites.

Dangerous new predators for adult amphibians in Latvia are invasive mammals that are becoming widespread in the country: the raccoon dog (*Nyctereutes procyonoides*) (29,200 individuals registered in 2014), the American mink (*Neovison vison*) (23,200 individuals registered in 2014), and probably the muskrat (*Ondatra zibethicus*) (5,500 individuals registered in 2014) (NeoGeo.lv 2014).

Research on the presence and distribution of local or exotic parasites in the *Pelophylax* kl. *esculentus* complex was carried out in Latvia in 2017. One-hundred-and-forty-one adult and juvenile specimens were collected for parasitological examination. The parasites found belong to the following systematic groups: Protozoa, Monogenea, Digenea, Nematoda and Acanthocephala. Fifty-five *Pelophylax* kl. *esculentus* complex tadpoles were examined. Three Trematoda species in their larval stage: *Alaria alata, Diplostomum* sp., *Opisthioglyophe ranae* and *Diplodiscus subclavatus* in its adult stage were detected in different organs of the tadpoles. One Nematoda, *Thelandros tba*, a common parasite of frog tadpoles, was observed in the intestine (Kirjusina *et al.* 2018a, 2018b).

One-hundred-and-thirty-five samples of *Pelophylax* kl. *esculentus* complex from 42 locations were checked for the presence of *Batrachochytrium dendrobatidis* (*Bd*) and *B. salamandrivorans* (*Bsal*) using qPCR in August 2017. *Bsal* was not detected in any sample. *Bd* infection was detected in 18 (43%) locations and in 53 (39%) samples. The infection intensity in the *P.* kl. *esculentus* complex reached a maximum of 143.0 genomic equivalents of zoospores (GE) in a water basin near Jurmala city (Kulikova *et al.* 2018).

Taking into account that *Xenopus laevis* and other tropical species of amphibians are widely sold in pet shops in Latvia and by Internet (Pupins and Pupina 2011a), it can be assumed that *Bd*, and other species of exotic parasites can be found in many populations of Latvian amphibians.

III. Declining species and species of special conservation concern

The Red Data Book (Andrusaitis 2003) is not the official nature protection document in Latvia; rather that role is taken by a popular book. The status of the species in the Red Data Book of Latvia does not directly influence practical actions towards species or their protection and can only serve for educational or illustrative purpose.

There is a special document in Latvia approved by the Cabinet of Ministers "Ministru kabineta noteikumi # 396" (Ministru kabineta noteikumi 2004) which contains a list of specially protected species in Attachment 1, including six species of amphibians. These are *Hyla arborea*, *Epidalea calamita*, *Bufotes viridis*, *Triturus cristatus*, *Bombina bombina* and *Pelobates fuscus*, all of which are species of special conservation concern in Latvia and in the European Union (Table 65.3).

In 2003 (Andrusaitis 2003) *Hyla arborea* was considered as extirpated in Latvia. However, in 1987 the Laboratory of Ecology, led by herpetoculturist Dr. Juris Zvirgzds-Zvirgzdiņš, was founded for remediation of this species at the Riga National Zoo. He considered widespread amelioration and a complete extirpation of beavers, *Castor fiber*, from Latvia to be the reasons for the decline of the treefrog (J. Zvirgzds-Zvirgzdiņš, personal communication). Using *Hyla arborea* that had been brought from Belarus for the formation of a breeding group and bringing 4,110 young individuals that had been bred in zooculture for reintroduction into nature in 1988–1992, the Riga National Zoo formed a stable population of *Hyla arborea* in Liepajas novads in southwestern Latvia (56° 30′ N) (Zvirgzds et al. 1995; Zvirgzds 2003). This population has been continuously increasing in numbers and extending its area of distribution. Now the area of expansion occupies Kurzeme and at least 1,000 km² in Kalvene, Vaiņode, Aizpute and Priekuli (Zvirgzds 2003). However, the present distribution of this species in Latvia does not mean that the danger of decline has disappeared. There was also an opinion that *Hyla arborea* had never inhabited Latvia and that it had been introduced into the country as an alien species (Caune 1992; A. Berzins, personal communication).

Epidalea calamita is a very rare species in Latvia, mainly inhabiting the coastal region of the Baltic Sea and its Gulf of Riga and the central part of Latvia (Caune 1992; Berzins 2008; Pupins and Pupina 2011a). Over the past few years it has been discovered in Dviete in the southeast of Latvia (E. Racinskis, personal communication). A. Berzins (2008) wrote that

> "… the Natterjack Toad inhabits mainly open, well-warmed landscapes with light, sandy soils. There it lives in sand dunes, glades of pine forests, sand and gravel quarries and meadows … … The reason for decline in the Natterjack Toad is complex and not fully understood, … and it reflects a continuous change in the cultural landscape over the past hundred years. The former distribution in Latvia included many sandy exposed coastal areas, but also inland sand or gravel pits with open ground and with oligotrophic and shallow breeding waters. The open exposed areas on poor and sandy

Table 65.3 The species of amphibians of special conservation concern and their status in Latvia (Latvian Environment, Geology and Meteorology Centre 2015; Ministru kabineta noteikumi 2004; Andrusaitis 2003).

Species	Category in Red Data Book of Latvia	Relevant EU legislation	Bern Convention	Protected species in Latvia (MKN #396)
Triturus cristatus	2 (vulnerable species)	HD II, IV	II	Yes
Bombina bombina	1 (endangered species)	HD II, IV	II	Yes
Hyla arborea	0 (extinct species)	HD IV	II	Yes
Pelobates fuscus	4 (undetermined species)	HD IV	II	Yes
Bufotes viridis	3 (rare species)	HD IV	II	Yes
Epidalea calamita	2 (vulnerable species)	HD IV	II	Yes

soils with a natural fluctuating water regime have become rare and the ground often becomes overgrown by dense vegetation, leading to changed microclimate …".

Bufotes viridis occurs throughout the whole territory of Latvia except for its northern part (Caune 1992; Pupins and Pupina 2011a). Nowadays the northern border of its current European distribution is partly located in Latvia; the populations of the species are tessellated, and their number is not large. It tends to live in anthropogenic landscapes and in the basins of large rivers in Latvia (Rimicans *et al.* 2010). The main reason for its decrease in number is the rapid overgrowing of terrestrial habitats and small ponds by bushes following the abandonment of unprofitable farms, thereby destroying habitat essential for this species. As an example, a large population of *Bufotes viridis*, which earlier had bred on the flood meadows used by local citizens for haymaking and as pasture for cows, disappeared in Niderkuni, Daugavpils, between 1987 and 2005 (A. Pupina and M. Pupins unpublished data).

Triturus cristatus is rare in Latvia; the species is tessellatedly distributed through the whole of the country (Caune 1992; Pupins and Pupina 2011a), also in the northern part; the populations are not large and are fragmented.

Bombina bombina is the rarest species of amphibian in Latvia. Until 2004 only two populations were known: at Islice (Bauska) and Ilgas (Silene), with up to ten vocalizing males in each population (Andrusaitis 2003). In 2004 A. Pupina, Ag. Pupina, and M. Pupins discovered a new population in Ainavas (Daugavpils region) (Pupins and Pupina 2006) and during the following years they came across populations in Demene, Medumi, Ilukste, Aizkraukle and Eglaine (Pupina and Pupins 2007; Pupins and Pupina 2011a).

All populations known in Latvia reside within 0–20 km from the southern and southeastern border of Latvia (Kuzmin *et al.* 2008; Pupina and Pupins 2008). The main reasons for the decline are the overgrowing of habitats by bushes and canes (Pupina *et al.* 2012); the occurrence of fish in ponds, especially the invasive *Perccottus glenii*; cleaning of soil reclamation canals; and destruction of dams and bodies of water produced by beavers. Two of the present authors (A. Pupina and M. Pupins) observed some cases of *Pelophylax* sp. actively preying upon juveniles of Bombina *bombina* (Figure 65.9). They also noticed an increase in some *Bombina bombina* populations and an

Fig. 65.9 *Pelophylax* sp. eating juvenile *Bombina bombina* (Demenes pagasts, Daugavpils novads; 55° 42′ 47.91″ N; 26° 29′ 54.88″ E). Photograph by Aija Pupina.

expansion of the species to the north, possibly influenced by climatic changes in Latvia (Pupina and Pupins 2010, 2013a).

Pelobates fuscus is spread throughout the territory of Latvia (Caune 1992; Pupins and Pupina 2011a); this species is inclined to live in mellow and sandy soils, including agricultural ones (Rimicans *et al.* 2010). The populations are not large and are tessellatedly expanded and fragmented, especially in the north of Latvia. The main reasons for the decrease in numbers are the overgrowing of meadows by bushes following cessation of haymaking, the overgrowing of breeding ponds by bulrushes, and the invasion of ponds by fishes.

Bombina bombina, Triturus cristatus and *Pelobates fuscus* often are sympatric in breeding ponds (Pupina and Pupins 2009). The main reasons for this species' decrease in numbers are (1) drainage of wetlands, (2) the overgrowing of breeding ponds by bushes, especially ponds in forests and bordering forests, and (3) invasion of ponds by fishes.

All of these species of amphibians are currently living in Latvia at the northern edge of the species' European distribution, or close to it (Caune 1992; Pupins and Pupina 2011a) and are vulnerable to climatic changes. The Latvian Climate Research Agency stated that

> "… during the 20th century the average air temperature in Latvia has risen by 1 degree. This increase is even more apparent in the years of the 21st century, which have already passed. During the past 100 years, there have been fluctuations in annual rainfall, which tended to rise from the beginning of the second half of the 20th century" (Latvian Biodiversity Clearing-House Mechanism 2015) (Figure 65.10).

Such changes of climate, namely increase of the temperature and humidity, can lead to changes both in the habitats of rare Latvian amphibians (e.g. a more rapid invasion of water by vegetation; changes in floral composition) and in distribution and number of amphibian species and its main invasive threat *Perccottus glenii* in Latvia, especially in those habitats on the northern border of the species' ranges (Nekrasova *et al.* 2018; Tytar *et al.* 2018).

IV. Conservation of amphibians in Latvia

All amphibians are protected in Latvia as an integral part of nature; there also are special measures for conservation of particularly protected species. The conservation of amphibians in Latvia is carried out on several levels: (1) the legislative level (adoption of laws, norms and requirements by government and competent organizations, aimed at amphibian conservation); (2) the planning level (development and approval of the Species Protection Plans for specially protected species of

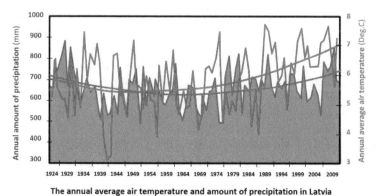

The annual average air temperature and amount of precipitation in Latvia

Fig. 65.10 The annual average air temperature and precipitation amount in Latvia (Latvian Environment, Geology and Meteorology Centre 2015).

amphibians in Latvia and of the Nature Conservation plans for specially protected natural reservations); (3) the level of territorial protection of nature (amphibian conservation on protected nature reserves); (4) the project-actions level (implementation of projects for conservation, financed by local governments and by the European Union); and (5) the level of particular specialists' and landowners' private actions. An official permit of the Nature Conservation Agency of Latvia is required for any action affecting particularly protected amphibians or their respective protected territories in Latvia. Such actions are assessed by a certified expert of the Nature Conservation Agency in the category named "Amphibians and Reptiles" or "Amphibians". There were five of these experts in Latvia in 2018 (Expert Register of Nature Conservation Agency 2018). In fact, all the amphibian conservation measures in Latvia described below were developed and implemented with the involvement of at least of one of these experts.

A. The legislative level

Latvia, as a member of the European Union, executes all its legislative norms and there are laws and regulations that are directly or indirectly involved with the conservation of nature, including amphibians. The most important Latvian laws are the Environmental Protection Law [Par vides aizsardzibu], Law On Specially Protected Nature Territories [Par ipasi aizsargajamam dabas teritorijam], General Regulations on Protection and Use of Specially Protected Nature Territories [Ipasi aizsargajamo dabas teritoriju visparejie aizsardzibas un izmantosanas noteikumi], Regulations on the Content and Elaboration Procedure of the Nature Protection Plan for Specially Protected Nature Territories [Noteikumi par ipaši aizsargajamas dabas teritorijas dabas aizsardzibas plana saturu un izstrades kartibu], Law on the Conservation of Species and Biotopes [Sugu un biotopu aizsardzibas likums], and others.

The Regulations on the List of the Specially Protected Species #396 [Noteikumi par ipasi aizsargajamo sugu un ierobezoti izmantojamo ipasi aizsargajamo sugu sarakstu Nr. 396] are the most important ones aimed particularly at conservation of declining amphibians in Latvia wherein an official list of six specially protected species is given: *Hyla arborea, Epidalea calamita, Bufotes viridis, Triturus cristatus, Bombina bombina* and *Pelobates fuscus* (Ministru kabineta noteikumi Nr. 396; 2004).

The Regulations for the Establishment, Conservation and Management of Micro-reserves #45 [Mikroliegumu izveidosanas, aizsardzibas un apsaimniekosanas noteikumi Nr. 45] are also important because they allow the development of micro-reserves for three species of the specially protected amphibians in Latvia: *Epidalea calamita, Bombina bombina* and *Triturus cristatus*.

B. The planning level

There are programmes and plans developed in Latvia that are indirectly or directly aimed at amphibian conservation. The most important of the general plans are the National Programme on Biological Diversity, the Latvian Environmental Policy Guidelines for 2014–2020.

A legislatively approved provision on the Plans for Management of Species exists in Latvia, i.e. Species Conservation Plan. Such plans are a national strategy for conservation of a particular species and describe the state of that species in Latvia, indicate the threats against it and its habitats, and offer measures for conservation of the species. In 2018, there were two plans approved for two amphibian species: *Bombina bombina* (Pupins and Pupina 2006) and *Epidalea calamita* (Berzins 2008). Such plans are updated every five to ten years.

There are many Nature Protection plans elaborated for particular specially protected territories in Latvia. Such plans also include measures aimed at conservation of specially protected amphibian species: as an example, *Bombina bombina* in microreserves Natura 2000 Ilgas, Islice, Katriniski and Strauti.

C. The level of territorial protection of nature

This level is related to the existence and operation of protected natural reserves. There are micro-reserves in Latvia that were founded for the conservation of specially protected declining amphibian species. The following micro-reserves and Natura 2000 territories were chosen for the conservation of *Bombina bombina*: Katriniski micro-reserve, Strauti micro-reserve (Demenes pagasts, Daugavpils novads) (Pupina and Pupins 2014b); Ilgas micro-reserve (Skrudalienas pagasts, Daugavpils novads); Īslīce micro-reserve (Īslīces pagasts, Bauskas novads). For conservation of *Epidalea calamita*: Puzes Reserve (Ventspils novads); Garkalnes Reserve (Dobeles novads); Karateru Reserve (Limbazu novads); Ainazi Reserve (Limbazu novads); Adazi Rezerve. *Hyla arborea*: in Embute pagasts, Liepajas novads (Berzins 2008). Other sympatric declining amphibian species often inhabit the same micro-reserves; for instance, *Pelobates fuscus* and *Triturus cristatus* occur in the micro-reserves of *Bombina bombina* (Pupina and Pupins 2009). Amphibian species inhabiting all the specially protected territories are protected according to the norms of those territories.

D. The project actions level

Particular projects for conservation of specially protected amphibian species and their habitats are implemented in Latvia with the aid of the EU European Commission and LIFE Programme; the Ministry of Environmental Protection and Regional Development and its structural units (Administration of Latvian Environmental Protection Fund, Nature Conservation Agency), local governments (Daugavpils city council) and organizations (e.g. Riga National Zoo, Latgales Zoo, Daugavpils University, Latvian Fund for Nature and Latgales Ecological Society).

Such projects can be international or national and usually foster complex actions: (1) conservation *in-situ*, (2) conservation *ex-situ*, (3) research (ecology, distribution, threats, habitats, herpetoculture), planning (Pupina and Pupins 2013b), and monitoring, (4) education of local inhabitants, and (5) dissemination of results. For example, three projects for conservation of rare amphibians have been implemented in Latvia and are co-financed by the European Commission and LIFE Programme:

The 2010–2014 LIFE project "Conservation of rare reptiles and amphibians in Latvia" (acronym: LIFE-HerpetoLatvia; LIFE09NAT/LV/000239) was co-financed by the European Commission: Country: Latvia. Target amphibian species: *Bombina bombina*. Two new microreserves and new Natura 2000 territories "Strauti" and "Katriniski" were created in Demene parish; 27 ponds were restored or created; more than 4,000 *Bombina bombina* juveniles were raised in amphibians' aquaculture (Figure 65.11); and released for population enforcement in 2013–2014 (Figure 65.12). The Project is very successful; both project sites are populated by *Bombina bombina*; also an additional site "Ilgas" which was improved for *Emys orbicularis* (Pupina and Pupins 2014a, 2014b).

The 2004–2009 LIFE project "Protection of habitats and species in Natura Park Razna" (LIFE04NAT/LV/000199) was co-financed by the European Commission. Country: Latvia. Target amphibian species: *Bombina bombina*. New ponds were created in Nature Park Razna and juvenile *Bombina bombina* were introduced to establish a new population.

The 2004–2008. LIFE project "*Bombina* in the Baltic Region – Management of fire-bellied toads in the Baltic region" (LIFE-Bombina number LIFE04NAT/DE/000028) was co-financed by the European Commission. Countries: Germany, Denmark, Sweden and Latvia. Target species: *Bombina bombina*. In two microreserves and Natura 2000 territories "Islice" and "Ilgas", new ponds were created and juveniles were released for population enforcement. The Project was very successful (Brockmuller and Drews 2009).

Many conservation projects and research on amphibians in Latvia have been executed with the financial support of the Latvian Environmental Protection Fund and Nature Conservation Agency.

Fig. 65.11 Breeding laboratory and recirculation system for amphibians in Rare Reptiles and Amphibians Breeding Centre (LIFE-HerpetoLatvia Project, Latgales Ecological Society, Ainavas, Daugavpils novds; 55° 50′ 65″ 5.65″ N; 26° 29′ 6.60″ E

Fig. 65.12 Releasing of *Bombina bombina* in restored pond where *Pelophylax sp.*, *Triturus cristatus*, and *Pelobates fuscus* also are breeding (LIFE-HerpetoLatvia Project, Katriniski microreserve, Demene, Daugavpils novads; 55° 43′ 10.56″ N; 26° 32′ 54.51″ E).

V. Monitoring of amphibians in Latvia

Latvia undertakes long-time monitoring and scientific analyses of amphibian population trends. The first methodology for a state-financed amphibian monitoring programme was developed in 2005 as part of a national biodiversity monitoring programme. It consisted of two major parts: Natura-2000 monitoring and whole-state or so-called "background" monitoring. Biodiversity of Natura-2000 sites were surveyed by formally approved monitoring methods during the EMERALD Project realized by the Latvian Fund for Nature in 2001–2004. Among other protected species, there were two Habitats Directive Annex-II amphibian species for which project datasheets were prepared: great crested newt (*Triturus cristatus*) and fire-bellied toads (*Bombina bombina*). *Triturus cristatus* was registered in seven territories, but *Bombina bombina* only in two. The information consisted mostly of presence-absence data with some evaluation of available habitats. The methodology for Natura 2000 was developed in 2007 by an environmental consulting company, Estonian, Latvian & Lithuanian Environment.

The methodology for "background" monitoring was approved in 2006 by a state agency: Latvian Environment, Geology and Meteorology Centre. Unlike monitoring by Natura 2000, the methods were applicable to all amphibian species. This monitoring was performed only once, in 2008, by the company Estonian, Latvian & Lithuanian Environment. It consisted of two types of surveys. The first was the counting of vocalizing males in spring in three repetitions on the same transect at intervals of two weeks. The other method was the setting of three-armed drift fences twice a season – in late spring and in autumn. In each case traps were set for a period of about one week during which they were checked three times. Trapped amphibians were removed and released, but not marked. Using a random stratified sampling method, a total of nine squares, each 1 x 1 km, in different places of Latvia were selected for monitoring. There was a transect ~0.5–1 km long for counting vocalizing males (for periods of one hour) and one drift-fence trap (four buried buckets and six metres of fence) in each. Results from vocalizing males and from the traps in spring generally were poor, but in autumn the results from traps were better, with up to 30 amphibians captured for each check of the traps. Only the most common species were recorded (*Lissotriton vulgaris, Bufo bufo, Rana arvalis, Rana temporaria, Pelophylax* sp.). Results are available in the form of a report accompanied by datasheets from the field on the Nature Conservation Agency's home site (http://biodiv.daba.gov.lv).

In 2013, a state-approved biodiversity monitoring programme and a methodology for monitoring amphibians were significantly reworked and expanded by the Latvian Fund for Nature during a state-financed project.

The State monitoring programme for amphibians began in 2015 and was finished in 2018. It consisted of a vocalizing anuran survey on 65 plots that covers the whole territory of the state and included monitoring of the great crested newt at 21 sites where this species was recorded in 1980–2010. A survey of vocalizing anurans indicated the presence of green frogs (*Pelophylax* sp.) in 100% of visited 5 x 5 km squares; common toad (*Bufo bufo*) – 93%; common frog (*Rana temporaria*) – 83%; moor frog (*Rana arvalis*) – 80%; spade-foot toad (*Pelobates fuscus*) – 13%; treefrog (*Hyla arborea*) – 10 %; and fire-bellied toad (*Bombina bombina*) – in 5%. Data from a survey by handnet of the great crested newt (*Triturus cristatus*) confirmed its presence in 60% of historical sites.

We have been conducting personal monitoring of *Bombina bombina* since 2000 in the previously known population at Ilgas, and since 2004 in newly discovered populations at Demene, Ainavas, Medumi, Bauska and Verdini. Methods of monitoring include the counting of vocalizing males and visual accounting, partly looking for tadpoles and eggs. The characteristics of ponds, the presence of fishes (especially the invasive *Perccottus glenii*) and the degree of overgrowing by vegetation were also estimated. We explored all known localities in Latvia 1–4 times per season until 2011; the localities have been selectively explored in 2012–2017 because of an increase in the number of localities found to contain this species (> 300). The results have been entered in the database and published (Pupina and Pupins 2008, 2010, 2012, 2013a; Pupina *et al.* 2008).

VI. Conclusions

Latvia, with its large quantity of humid habitats and small ponds, non-freezing bodies of water and other places for overwintering, undeveloped agricultural industry and decreasing human population, generally corresponds to the needs of the amphibian species that inhabit it. However, Latvia is a northern European country and a half of all the Latvian amphibian species – *Triturus cristatus, Bombina bombina, Hyla arborea, Bufotes viridis, Epidalea calamita, Pelobates fuscus, Pelophylax* sp. – reach the northern border of their European distribution in or near Latvia. Therefore, their populations in Latvia are few in number and are fragmented. These species have special require-ments for optimizing their habitats in Latvia, i.e. increasing the permanence and luminance of non-freezing ponds used by amphibians for overwintering. The progressive overgrowing by bushes, shading, and degradations of such habitats contribute to the decline of the amphibian populations. The increasing threats of anthropogenic origin – mortality on roads, invasive species of fishes, invasive mammalian predators, invasive freshwater turtles and new parasites may lead to the decline of the few amphibian species inhabiting Latvia, and accordingly their European area of distribution will decrease. The climatic changes in Latvia can have both positive and negative meaning for Latvian amphibians. Because of climatic change, research is urgently needed on (1) the peculiarities of the ecology and distribution of declining amphibians on the northern border of Latvia; (2) optimization of biotopes and enhancement of the endangered populations using animals from zooculture; and (3) performance of the necessary measures aimed at protecting limited populations and at uniting them into metapopulations by establishing population bridges, including trans-border ones, for making genetic contact with stronger southern populations.

VII. Acknowledgements

Conservation of *Bombina bombina* in 2010–2014 was executed with the support of LIFE Project LIFE09NAT/LV/000239 "Conservation of rare reptiles and amphibians in Latvia" co-financed by the European Commission. The research in aquaculture was supported by the Rural Support Service of Latvia, Project 16-00-F02201-000002, "Innovation". The research on modelling was supported by the National Research Program EVIDEnT, No 4.6. "Freshwaters Ecosystems Services and Biodiversity", Project LHEI-2015-19. The article was prepared and researches of amphibians in Latvia have been executed with the financial support of the Latvian Environmental Protection Fund.

We thank the Institute of Life Sciences and Technologies of Daugavpils University, Daugavpils City Council, Latgales Zoo, Latgales Ecological Society, Nature Conservation Agency, Ministry of Environment and Regional Development of Latvia, and the administration of the Latvian Envi-ronmental Protection Fund for support of amphibian conservation in Latvia. We thank the editors of the book series Amphibian Biology for their kind help, reviewing and suggestions. We mourn the loss of our colleague and co-author, leading Latvian batrachologist Dr. Aija Pupina and are grateful for her highly valued cooperation in research and conservation of amphibians in Latvia.

VIII. References

Andrusaitis, G. (ed), 2003. *Red Data Book of Latvia*. University of Latvia, Riga.

Berzins, A., 2008. Smilsu krupja *Bufo calamita* (Laurenti, 1768) sugas aizsardzibas plans Latvija. [Species conservation plan for *Bufo calamita* (Laurenti, 1768) in Latvia]. Dabas aizsardzibas parvalde, Riga [in Latvian].

Brockmuller, N. and Drews, H., 2009. *Management of fire-bellied toad populations in the Baltic region*. Stiftung Naturschutz Schleswig-Holstein.

Caune, I., 1992. *Latvijas abinieki un rapuli*. [Latvian amphibians and reptiles]. Gandrs, Riga [in Latvian].

Central Statistical Bureau Republic of Latvia. Geographical Data – Key Indicators. http://www.csb.gov.lv/en/statistikas-temas/natural-resources-key-indicators-30500.html. Downloaded 14 February 2015.

Expert Register of Nature Conservation Agency. http://www.daba.gov.lv/public/lat/dati1/dabas_ekspertu_registrs/#Registrs. Downloaded 12 May 2018.

Kirjusina, M., Gravele, E., Mezaraupe, L. and Pupins, M., 2018a. First data of green frog *Pelophylax esculentus* complex (Anura: Ranidae) tadpoles parasites in Latvia. *Abstract Book of 2nd International Conference "Smart Bio". 2018.05.03.–05.05.*

Kirjusina, M., Gravele, E., Mezaraupe, L. and Pupins, M., 2018b. The pilot study of green frog *Pelophylax esculentus* complex (Anura: Ranidae) parasites in Latvia. *Book of abstracts. 60th International Scientific Conference of Daugavpils University. 2018.04.26.–2018.04.27.*

Kulikova, E., Balaz, V., Pupina, A., Pupins, M. and Ceirans, A., 2018. First record of *Batrachochytrium dendrobatidis* in wild populations of *Pelophylax esculentus* complex in Latvia. *Book of abstracts. 8th European Pond Conservation Network Workshop. Museu de la Mediterrania. 2018.05.21.–25.*

Kuzmin, S.L., Pupina, A., Pupins, M. and Trakimas, G., 2008. Northern border of the distribution of the red-bellied toad *Bombina bombina*. *Zeitshrift fur Feldherpetologie* **15**: 215–28.

Latvian Biodiversity Clearing-House Mehanism. http://biodiv.lvgma.gov.lv/cooperation/fol288846/fol297026. Downloaded 14 February 2015.

Latvian Environment, Geology and Meteorology Centre http://www.meteo.lv/en/lapas/environment/climate-change/climate-of-latvia/climat-latvia?id=1471&nid=660. Downloaded 14 February 2015.

Ministru kabineta noteikumi Nr. 396, 2004. Noteikumi par ipasi aizsargajamo sugu un ierobezoti izmantojamo ipasi aizsargajamo sugu sarakstu. *Latvijas Vestnesis* **41** 3/41 7.

Nekrasova, O., Tytar, V., Pupina, A. and Pupins, M., 2018. Distribution modeling of the Fire-bellied toad, *Bombina bombina*, as a tool for conservation planning under global climate change. *Book of abstracts. 8th European Pond Conservation Network Workshop. 2018.05.21.–25.*

NeoGeo.lv. http://neogeo.lv/?p=20108. Downloaded 14 February 2015.

Pupina, A. and Pupins, M., 2007. A new *Bombina bombina* L. population "Demene" in Latvia, Daugavpils area. *Acta Universitatis Latviensis* **273**: 47–52.

Pupina, A. and Pupins, M., 2008. The new data on distribution, biotopes and situation of populations of *Bombina bombina* in the southeast part of Latvia. *Acta Biologica Universitatis Daugavpiliensis* **8**: 67–73.

Pupina, A. and Pupins, M., 2009. Comparative analysis of biotopes and reproductive-ecological manifestations of *Bombina bombina* (Linnaeus, 1761) in Latvia. *Acta Biologica Universitatis Daugavpiliensis* **9**: 121–30.

Pupina, A. and Pupins, M., 2010. Dinamika chislennosti zherlanki krasnobryuhoi (*Bombina bombina* L. 1761) na severnoy granice areala v Latvii. [Dynamics of number of Fire-bellied toad (*Bombina bombina* L. 1761) at north edge of its distribution in Latvia]. *Vestnik Mozyrskaga Dzyarzhavnaga Universiteta imya I.P. Shamyakina* **26**: 30–4 [in Russian].

Pupina, A. and Pupins, M., 2012. Rasprostranenie krasnobryuhoi zherlanki *Bombina bombina* (Linnaeus 1761) v Latvii. [Distribution of Fire-bellied toad *Bombina bombina* (Linnaeus 1761) in Latvia]. *The problems of Herpetology. Proceedings of the 5th Congress of the Alexander M. Nikolsky Herpetological Society. 24–27 September 2012*. Minsk, Belarus: 265–8 [in Russian].

Pupina, A. and Pupins, M., 2013a. Fire-bellied toads *Bombina bombina* L. (Anura: Bombinatoridae) populations expansion in 2010 in Latvia. *Proceedings of the 9th National Congress of Italian National Herpetological Society, 26–30.09.2012*: 338–41.

Pupina, A. and Pupins, M., 2013b. LIFE-HerpetoLatvia: Population management corrected plan for the Fire-bellied Toad (*Bombina bombina*) population in Demenes pagasts (Daugavpils novads, Latvia). LIFE-HerpetoLatvia, Daugavpils.

Pupina, A. and Pupins, M., 2014a. LIFE-HerpetoLatvia: Results of monitoring of Fire-bellied Toad (*Bombina bombina*) habitat improvement actions in Demene (Latvia). Action E.3. LIFE-HerpetoLatvia, Daugavpils.

Pupina, A. and Pupins, M., 2014b. Project LIFE-HerpetoLatvia: first results on conservation of *Bombina bombina* in Latvia. *Herpetological Facts Journal. Supplement of the 2nd International workshop–conference: "Research and conservation of European herpetofauna and its environment: Bombina bombina, Emys orbicularis, and Coronella austriaca". 14–15.08.2014* **1**: 76–84.

Pupina, A., Pupins, M. and Berzins, A., 2008. New data on the distribution of *Bombina bombina* in Latvia on the northern edge of its area. *Biologia plazow i gadow – ochrona herpetofauny. IX Ogolnopolska Konferencja Herpetologiczna*: 194–8.

Pupina, A., Pupins, M., Ivanova, T. and Kotane, L., 2012. LIFE-HerpetoLatvia: Results of preliminary study of the Fire-bellied Toad (*Bombina bombina*) population Demene (Demenes pagasts, Daugavpils novads, Latvia). LIFE-HerpetoLatvia, Daugavpils.

Pupina, A., Pupins, M., Skute, A., Pupina A. and Karklins, A., 2015. The distribution of the invasive fish amur sleeper, rotan *Perccottus glenii* Dybowski, 1877 (Osteichthyes, Odontobutidae), in Latvia. *Acta Biologica Universitatis Daugavpiliensis* **15 (2)**: 329–41.

Pupins, M. and Pupina, A., 2006. "Sarkanvedera ugunskrupja *Bombina bombina* (Linnaeus, 1761) sugas aizsardzibas plans Latvija". ["Species conservation plan for *Bombina bombina* (Linnaeus, 1761) in Latvia"]. Dabas aizsardzibas parvalde, Riga [in Latvian].

Pupins, M. and Pupina, A., 2007. Functions of beavers *Castor fiber* L. in preservation of a rare species of Latvia *Bombina bombina* L. in a southeast part of Latvia. In: *Proceedings of 2nd International scientific Conference "Conservation of animals and hunting management in Russia"*: 67–70 [in Russian].

Pupins, M. and Pupina, A., 2011a. "Latvijas pieaugušo abinieku sugu lauku noteicējs". ["Field guide to adult Amphibians of Latvia"]. Daugavpils Universitate, Akademiskais apgads Saule, Daugavpils [in Latvian].

Pupins, M. and Pupina, A., 2011b. First records of 5 allochthonous species and subspecies of Turtles (*Trachemys scripta troostii, Mauremys caspica, Mauremys rivulata, Pelodiscus maackii, Testudo horsfieldii*) and new records of subspecies *Trachemys scripta elegans* in Latvia. *Management of Biological Invasions* **2**: 69–81.

Pupins, M. and Pupina, A., 2012a. Invasive fish *Perccottus glenii* in biotopes of *Bombina bombina* in Latvia on the north edge of the fire-bellied toad's distribution. *Acta Biologica Universitatis Daugavpiliensis* Suppl. **3**: 82–90.

Pupins, M. and Pupina, A., 2012b. Findings of *Emys orbicularis* (Linnaeus 1758) in salmonid lakes in Latvia. *Acta Biologica Universitatis Daugavpiliensis* Suppl. **3**: 91–93.

Pupins, M. and Pupina, A., 2012c. *Trachemys scripta elegans*, potential competitor of *Emys orbicularis*, first successful wintering in natural climatic conditions in Latvia in an experiment. *Herpetological Information. Materials of the 27th Conference of Czech Herpetological Society. 18.05. – 20.05.2012*. **11**: 15–16.

Pupins, M., 2007. First report on recording of the invasive species *Trachemys scripta elegans*, a potential competitor of *Emys orbicularis* in Latvia. *Acta Universitatis Latviensis* **273**: 37–46.

Rimicans, A., Pupins, M. and Pupina, A., 2010. Amphibians and reptiles of Daugavpils city territory (Latvia, Latgale). *Biologia plazow i gadow-ochrona herpetofauny*: 110–3.

Silins, J. and Lamsters, V., 1934. *Latvijas rapuli un abinieki*. [Reptiles and Amphibians of Latvia]. Rapa, Riga [in Latvian].

The IUCN Red List of Threatened Species. Version 2014.3. www.iucnredlist.org. Downloaded 14 February 2015.

Tytar, V., Nekrasova, O., Pupina, A., Pupins, M. and Marushchak, O., 2018. Species distribution modelling as a proactive tool for long-term planning of management of the Fire-bellied toad *Bombina bombina* (Linnaeus, 1761) and its main invasive threat *Perccottus glenii* (Dybowski, 1877) in Latvia under global climate change. *Abstract Book of 2nd International Conference "Smart Bio". 2018.05.03.– 05.05.*

Zvirgzds, J., 2003. Reintroduction of the Commen tree frog *Hyla arborea* in Latvia. In *Red Data Book of Latvia*, ed G. Andrusaitis. University of Latvia, Riga.

Zvirgzds, J., Stasuls, M. and Vilnitis, V., 1995. Reintroductions of the European tree frog (*Hyla arborea*) in Latvia. *Memoranda Societatis pro Fauna et Flora Fennica* **71**: 139–42.

66 Amphibian declines and conservation in Lithuania

Giedrius Trakimas, Jolanta Rimšaitė and John W. Wilkinson

Abbreviations and acronyms used in the text and references

IUCN	*International Union for the Conservation of Nature*
LRL	*Lithuanian Red List*
LRV	*Lietuvos Respublikos Vyriausybė [Government of the Republic of Lithuania]*
SFS	*State Forest Service*

I. Introduction

Lithuania is a country on the eastern shore of the Baltic Sea with a total area of 45,300 km². The terrain is predominantly flat; lowlands (< 100 m a.s.l.) occupy a considerable part of the land area and only 20% of the country consists of hilly uplands. Most of these uplands are located in the east (with Aukštojas Hill being the highest point in Lithuania at 293.8 m a.s.l.), with the exception of the insular Žemaitija upland in the west (Guobytė and Satkūnas 2011). The climate is transitional between maritime and continental. The mean annual temperature in Lithuania is about 6–7°C, the highest mean temperature is on the coast (7.4–7.6°C) and decreases towards the east to 5.8°C. A closed 6°C isotherm encircles the Žemaitija upland. Mean annual precipitation is 675 mm with a range from about 600 mm in the central lowland to about 900 mm on the windward side of the Žemaitija upland; 60–66% of precipitation falls during the warm period (Bukantis 2013).

    Once heavily forested, Lithuania's area now consists of only 33.4% woodland, primarily coniferous stands (comprising 56% of the forested area) and deciduous forests (40.4%) (SFS 2015). The forests are highly fragmented but influence ecological processes in most landscapes; relatively less fragmented forest exists in the south-eastern part of the country (Kucas *et al.* 2011). The area covered by forest is steadily increasing at the expense of agricultural land, which currently comprises

53.4% of the total area (SFS 2015). There are many ponds and lakes and the density of the hydrographic network in Lithuania is 1.18 km/km^2. Regulated water courses constitute 82.6% and those with natural beds just 17.4% (Gailiušis *et al.* 2001).

There are 13 species of amphibians (11 anurans and two caudates) in Lithuania. This represents about 18% of the 74 indigenous species of Europe (Speybroeck *et al.* 2010). All amphibians found in Lithuania have extensive ranges in Europe and have successfully recolonized the current area following deglaciation. These species generally possess traits that indicate a potential for rapid expansion of range (Trakimas *et al.* 2016).

II. Distribution of amphibians in Lithuania

Records of the distribution of amphibians in Lithuania have been summarized in a national atlas (Balčiauskas *et al.* 1999). Furthermore, three local atlases (see Malinauskas *et al.* 1998, Malinauskas 2000, 2001) also have been published. Further investigations are required for complete accounts of distribution, and to specify the distributional patterns at different scales.

A. Caudata

Lissotriton vulgaris is widespread and common (Balčiauskas *et al.* 1999), being found all over Lithuania; it is much more abundant than *Triturus cristatus*. The latter species is also widespread throughout Lithuania, but its distribution is uneven. It is common in the south and southeast, although data are lacking on its distribution and, especially, on its abundance.

According to available data, the total population of great crested newts in Lithuania may be 3,000–5,000 adult individuals (Rašomavičius 2007). Recently, these newts have disappeared from some places but three to five new localities are found every year (Rašomavičius 2007). The abundance of great crested newts in Lithuania is several times lower than that of smooth newts, which have the same breeding habitats. In addition, the great crested newt is more demanding in its breeding and terrestrial habitat preferences than is the smooth newt and many other Lithuanian amphibians (Balčiauskas *et al.* 1999; Bastytė 2009).

B. Widespread anurans

Bufo bufo is the commonest and most widespread of Lithuania's toads. The species occurs at a density of two to nine individuals per ha in meadows and four to 16 individuals per ha in mixed forests, but it is most abundant at interfaces with cultivated fields (16 to 50 individuals per ha) (Balčiauskas *et al.* 1999). Šireika and Stašaitis (1999) also found that, in Aukštaitija National Park, common toads were most abundant in cultivated areas (37.7 ± 0.90 individuals per 0.1 ha), as well as in fir groves (19.7 ± 2.34 individuals per 0.1 ha). *Bufotes viridis*, on the other hand, is not common but is continuously distributed across the country. It is commonest in the south and southeast (Balčiauskas *et al.* 1999). *Epidalea calamita* is also widespread, being commonest in central Lithuania, although rare in other parts of the country. Densities of up to 28–100 individuals per ha have been recorded in the coastal zone, as compared to 3–27 individuals per ha in other parts of Lithuania (Balčiauskas *et al.* 1999).

Both species of brown frog (*Rana* spp.) are common and widespread. According to recent data (Trakimas 2008), the moor frog (*Rana arvalis*) can be considered as rather continuously distributed except in the northern part of Lithuania where it appears to be patchier. Densities can be high, with up to 180 – 220 individuals per ha (Balčiauskas *et al.* 1999) or 20 ± 2.13 individuals per 0.1 ha (Šireika and Stašaitis 1999; Aukštaitija National Park) in upland bogs. In the Kurtuvėnai Regional Park (northwest Lithuania), the moor frog was reported as being common in river valleys (Trakimas 1999). *Rana temporaria* is the most numerous amphibian in Lithuania, and is found in a variety of habitats. Densities have been reported of up to 350 to 750 individuals per ha within wet deciduous forests and even more abundant at their edges (Balčiauskas *et al.* 1999).

Of the green frogs, *Pelophylax lessonae* and *P.* kl. *esculentus* are both common and widespread (Balčiauskas *et al.* 1999). The former species prefers small bodies of water whereas the latter occurs in a variety of habitats. *Pelophylax ridibundus* is more locally distributed, being found most often near large rivers. It is commonest in Kuršiai Lagoon, the Nemunas river delta, and the Kaunas reservoir (Balčiauskas *et al.* 1999).

C. Less common anurans

Pelobates fuscus is local and uncommon in Lithuania, being found mostly in the west, south and southeast, with single records in other areas (Balčiauskas *et al.* 1999). Its habitats include sand pits, pine-woods and deciduous forests. There is still a lack of data on the Lithuanian distribution of this species. *Bombina bombina* is also distributed unevenly in the country. It occurs most frequently in the south and east (Balčiauskas *et al.* 1999) and is very rare in the west and north, although its distribution extends into southern Latvia (Kuzmin *et al.* 2008). Records from some decades ago of *B. bombina* in Lithuania's western municipalities (Mažeikiai, Rietavas) have not been confirmed by more recent studies (see Balčiauskas *et al.* 1999 and Malinauskas 2001).

The European treefrog, *Hyla arborea*, was considered to be extirpated in Lithuania but was rediscovered in the south in 1988. It remains rare and locally distributed, inhabiting wet shrubby meadows near the Nemunas river (Balčiauskas *et al.* 1999). In 2013, the species was found in Skuodas district, in the northwest of Lithuania (Raudonytė J. personal communication). It is likely that the Skuodas treefrogs have dispersed from the reintroduction sites in Latvia.

III. Threats

The major threats to Lithuanian amphibians are considered to be the destruction and alteration of habitats (e.g. draining of wetlands) (Gaižauskienė 1981), pollution, mortality on roads (Simanavičienė 2008; Bartaškaitė 2017), and increased fragmentation of habitat by networks of roads (Bartaškaitė 2017). However, detailed studies on the effects of threats to amphibian populations in Lithuania are lacking.

A land-reclamation campaign was initiated in the first half of the twentieth century in Lithuania, and large-scale drainage continued throughout that century, the total area drained being about three million ha (Karlsson and Ryden 2012). The mires and large floodplains of unregulated rivers have disappeared and were replaced by arable land and grassland (Karlsson and Ryden 2012).

A. *Epidalea calamita*

The main threats to this pioneer species are the drainage of land and intensive agriculture (Jankevičius *et al.* 1981; Rašomavičius 2007) but these are exacerbated by pollution and direct destruction of breeding habitats. Breeding sites with sandy substrates are being lost through draining and reforestation (Rašomavičius 2007).

B. *Triturus cristatus*

The breeding sites of this species are threatened by the introduction of fish into small ponds, as well as pollution of water and drainage (Balevičius 1992; Rašomavičius 2007). Qualitative changes, such as the deepening and clearance of vegetation from bodies of water also have negative impacts. This species exists in a metapopulation structure, and intensive agricultural practices, including the use of synthetic fertilizers, may prevent successful dispersal between patches of habitat.

C. *Hyla arborea*

The greatest threat to treefrogs in Lithuania is destruction and degradation of breeding habitats (Rašomavičius 2007), and also isolation of breeding sites and feeding habitats. Intensive agriculture and pollution of breeding ponds and land habitats have strong negative impacts. Population densities depend on meteorological conditions and the hydrological conditions of bodies of water.

D. *Bombina bombina*

Balevičius (1992) stated that suitable breeding habitats for *B. bombina* are decreasing due to pollution and to drainage. Some local Lithuanian populations that occur in fishery ponds have been lost or diminished due to changes in fishery techniques; others suffer mortality during seasonal (spring and autumn) migrations (Rašomavičius 2007).

E. *Bufotes viridis*

Transition of habitats to late successional stages (overgrown by grass, scrub, and forests) negatively impacts this species. There has also been a reduction in the number of suitable breeding sites, including those impacted by usage of pesticides and herbicides, as well as by the introduction of fish (Rašomavičius 2007). The fragmentation of habitat threatens the integrity of migration routes (*ibid.*).

IV. Conservation status

The overall conservation status of Lithuanian amphibian species is relatively good; none is considered threatened at the European level (del Carpio *et al.* 2013) and all are classified as Least Concern. Nationally, however, five species of amphibians are listed in the Lithuanian Red List (LRL): *Triturus cristatus, Bombina bombina, Hyla arborea, Epidalea calamita*, and *Bufotes viridis* (Rašomavičius 2007). The natterjack toad (*E. calamita*) was the first species of amphibian included in the LRL in 1976 for reasons of being both rare and in decline (Jankevičius *et al.* 1981). This sparked an interest in the toad and many new localities have since been reported (e.g. Berzinis and Margis 1987; Malinauskas and Mikutavičius 1993). Consequently, the species was considered as "restored" in later editions of the LRL (Balevičius 1992; Rašomavičius 2007). *Bombina bombina* was included in the LRL in 1989 but is also currently listed as restored (Rašomavičius 2007). *Triturus cristatus* was included in the LRL in 1991, *H. arborea* in 2000, and *B. viridis* in 2007. Thus, out of 13 amphibian taxa reported in Lithuania, one is rare (*H. arborea*), two are intermediate (*T. cristatus, B. viridis*) and two are considered as restored (*B. bombina, E. calamita*).

V. Conservation measures and monitoring

A. Special conservation areas for amphibians

A national herpetological reserve for the conservation of *Hyla arborea* was established in Baltoji Ančia in 1997, with an area of 13 ha (LRV 1998); later the area of the reserve was expanded to 15.7 ha. Thirty-seven areas of the Natura 2000 European ecological network are designated in Lithuania for conservation of habitats for rare amphibians. *Bombina bombina* is protected in 25 of these and *Triturus cristatus* is protected in 26 (there are 14 areas where both species are protected).

In 1988, the first municipal reserve for conservation of *Epidalea calamita* was established in Žasliai (Kaišiadorys district) (Malinauskas *et al.* 1998).

B. Projects

Since 2005, the Lithuanian Fund for Nature, with partners, has carried out projects including actions for active amphibian conservation in South Lithuania. During the LIFE Plus Nature project (LIFE09 NAT/LT/000581), an ecological network for *Triturus cristatus, Bombina bombina, Pelobates fuscus, Hyla arborea, Epidalea calamita, Bufotes viridis, Rana arvalis* and *Pelophylax lessonae* was created by restoring aquatic and terrestrial habitats, designating protected areas, and the head-starting of juvenile froglets of *Hyla arborea*. During the project, 164 ponds were dug in seven areas and 52 ponds were restored (Bastytė 2014). Also, many amphibian hibernacula were created in South Lithuania during a previous LIFE Natura project (LIFE05 NAT/LT/000094) in 2005–2009 (Adrados and Schneeweiss 2009).

In the latter part of the twentieth century, many local initiatives were designed to reduce the mortality of amphibians on roads. Usually, temporary fences were built at the sides of roads during the breeding season. Temporary mitigations of amphibian mortality on roads (temporary fences erected during the breeding season) were built in Varniai Regional Park beginning in 2000. Later, similar initiatives were carried out in Pavilniai, Verkiai, Veisiejai and Kurtuvėnai regional parks and in Kamanos State Nature Reserve. From 2012 onwards, several permanent mitigating systems, with underpasses (tunnels) and barrier walls were built in South and East Lithuania (Lopeta 2017).

C. Monitoring

Since 2007, the monitoring of amphibian species of European importance (i.e. *Bombina bombina* and *Triturus cristatus*) has been carried out three times (2007, 2010, and 2013–2014) at Natura 2000 sites (www.gamta.lt). Monitoring of *B. bombina* was implemented at eight Natura 2000 sites and in two fishery-pond systems in 2007, at seven Natura 2000 sites in 2010, and at 23 Natura 2000 sites in 2013–2014. *Triturus cristatus* was monitored at six Natura 2000 sites in 2007, at five in 2010, and at 22 in 2013–2014.

Bombina bombina were found in ~90% of monitored Natura 2000 sites during the amphibian monitoring in 2013–2014. During that monitoring, several negative factors affecting *B. bombina*'s habitats were observed: presence of fish (in 36% of aquatic habitats sampled) and fragmentation and isolation of habitat (33% and 24%, respectively). The conservation status of *B. bombina* was considered as unfavourable – inadequate.

During monitoring in 2013–2014, *T. cristatus* was found at 50% of monitored Natura 2000 sites. The following negative environmental factors were registered: presence of fish (63% of aquatic habitats), large populations of water frogs (57%), isolation of habitats (42%), fragmentation of habitats (44%). The conservation status of *T. cristatus* was considered as unfavourable – inadequate (Balčiauskas 2015).

VI. Acknowledgements

We thank V. Naruševičius, M. Simanavičienė, U. Trakimaitė, and N. Zableckis for their assistance.

VII. References

Adrados L.C. and Schneeweiss N., 2009. Save Europe's Oldest Reptile and Amphibians. LIFE Nature project (LIFE05 NAT/LT/000094) layman's report. Vija, Druskininkai.

Balčiauskas, L., Trakimas, G., Juškaitis, R., Ulevičius, A. and Balčiauskienė, L., 1999. *Atlas of Lithuanian Mammals, Amphibians and Reptiles*. Akstis Publ. Vilnius.

Balčiauskas, L., 2015. Europos Bendrijos svarbos rūšių būklės ir ivazinių ugalų ir gyvūnų rūšių tyrimo atlikimo paslaugos ataskaita. Vilnius. [Report of the research services on the monitoring of European Community important and invasive animals and insects species].

Balevičius, K. (ed.), 1992. *Lietuvos raudonoji knyga* [Red Data Book of Lithuania]. Lietuvos Respublikos aplinkos apsaugos departamentas, Vilnius.

Bartaškaitė J., 2017. Varliagyvių žuvimo keliuose priežastys. [Causes of Amphibian Road Mortality]. Bachelor thesis, Vilnius University.

Bastytė, D., 2009. Priešmetamorfinių *Triturus cristatus* ir *T. vulgaris* aptinkamumo ir gausumo priklausomybė nuo aplinkos veiksnių Pietvakarių Lietuvoje ir šalia esančiose Lenkijos teritorijose (Master thesis, Vilnius University).

Bastytė, D., 2014. Development of a Pilot Ecological Network in South Lithuania. LIFE Plus Nature Project (LIFE09 NAT/LT/000581) layman's report. LFN, Vilnius.

Berzinis, A. and Margis, G., 1987. Nendrinės rupūžės [Natterjack toads]. *Mūsų gamta* **10**: 22.

Bukantis, A., 2013. Klimatas. In *Lietuvos gamtinė geografija* [Natural Geography of Lithuania] ed. M. Eidukevičienė. Klaipėdos universiteto leidykla, Klaipėda [in Lithuanian].

EB svarbos gyvūnų rūšių (lūšies, šikšnosparnių, varliagyvių) monitoringas/Aplinkos apsaugos agentūra, from www.gamta.lt [Monitoring of EC species, The Environmental Protection Agency].

Gailiušis, B., Jablonskis, J. and Kovalenkovienė, M., 2001. *Lietuvos upės. Hidrografija ir nuotėkis.* [Lithuanian Rivers. Hydrography and runoff.] Lietuvos energetikos institutas, Kaunas [in Lithuanian].

del Carpio, P.A., Sánchez S., Nieto A. and Bilz M., 2013. *Lithuania's biodiversity at risk. A call for action.* Publications Office of the European Union, Luxembourg.

Gaižauskienė, J., 1981. *Susipažinkite: Varliagyviai ir Ropliai* [Be Acquainted with Amphibians and Reptiles]. Mokslas, Vilnius.

Guobytė, R. and Satkūnas, J., 2011. Pleistocene Glaciations in Lithuania. *Development in Quaternary Science* **15**: 231–46.

Jankevičius, K., Zajančkauskas, P., Balevičius, K., Kazlauskas, R., Lekavičius, A., Logminas, V., Mačionis, A., Sukackas, V. and Tursa, G. (eds.), 1981. *Lietuvos TSR raudonoji knyga* [Red Data Book of Lithuanian SSR]. Mokslas, Vilnius.

Lietuvos Respublikos Vyriausybė (LRV), 1998. Dėl naujų draustinių įsteigimo ir draustinių sąrašų patvirtinimo [On Establishment of New Reserves and Approval of Lists of Reserves]. Nutarimas Nr.1486 (1997–12–29). *Valstybės žinios*, Nr. 1–9.

Lopeta V., 2017. Varlianešio darbams pasibaigus [After frogs' carrying works are finished]. Retrieved 12 December 2017, from http://www.krpd.lt.

Karlsson, I. and Ryden, L. (eds.), 2012. *Rural Development and Land Use.* Baltic University Press, Uppsala.

Kučas, A., Trakimas, G., Balčiauskas, L. and Vaitkus, G., 2011. Multi-scale analysis of forest fragmentation in Lithuania. *Baltic Forestry* **17**: 128–35.

Kuzmin, S.L., Pupina, A., Pupins, M. and Trakimas, G., 2008. Northern border of the distribution of the Red-bellied toad (*Bombina bombina*). *Zeitschrift für Feldherpetologie* **15**: 215–28.

Malinauskas, V. and Mikutavičius D., 1993. Nendrinės rupūžės (*Bufo calamita*) naujos radimvietės Lietuvoje [New localities of Natterjack toad (*Bufo calamita*) in Lithuania]. *Raudoni lapai* **1**: 28–30.

Malinauskas, V., Juškys, K., Grėsius R. and Matas, V., 1998. Varliagyviai ir ropliai

[Amphibians and Reptiles]. In: Malinauskas, V. (ed.): *Kaišiadorių rajono gyvūnijos atlasas (1986–1997)*. Gamtos pasaulis, Vilnius.

Malinauskas, V. (2000): Varliagyviai ir ropliai [Amphibians and Reptiles]. In: Malinauskas, V. (ed.): *Rokiškio rajono gyvūnijos atlasas (1995–1999)*. Daigai, Vilnius.

Malinauskas, V. (2001): Varliagyviai ir ropliai [Amphibians and Reptiles]. In: Malinauskas, V. (ed.): *Mažeikių rajono gyvūnijos atlasas (1990–2000)*. Piko valanda, Marijampolė.

Rašomavičius, V. (ed.), 2007. *Lietuvos raudonoji knyga* [Red Data Book of Lithuania] Lututė, Kaunas.

SFS 2015. *Lithuanian statistical yearbook of forestry, 2015*. Lututė, Kaunas.

Simanavičienė, M., 2008. Varliagyvių mirtingumas Pavilnių regioninio parko keliuose [Amphibian Road Mortality at the Pavilniai Regional Park]. Bachelor thesis, Vilnius University.

Speybroeck, J., Beukema, W. and Crochet, P.A., 2010. A tentative species list of the European herpetofauna (Amphibia and Reptilia) – an update. *Zootaxa* **2492**, 1–27.

Šireika, E. and Stašaitis, J., 1999. Abundance and Distribution of Amphibians in Aukštaitija National Park. *Acta Zoologica Lithuanica* **9 (3)**: 91–5.

Trakimas, G., 1999. Amphibian species diversity in Kurtuvėnai Regional Park. *Acta Zoologica Lituanica* **9**: 86–90.

Trakimas, G., 2008. The moor frog (*Rana arvalis*) in Lithuania: distribution and status. *Zeitschrift für Feldherpetologie* **13** (Suppl.): 207–10.

Trakimas, G., Whittaker, R.J. and Borregaard, M.K., 2016. Do biological traits drive geographical patterns in European amphibians? *Global Ecology and Biogeography* **25**: 1228–38.

67 Decline and conservation of amphibians in Denmark

Kåre Fog, Lars Christian Adrados, Andreas Andersen, Lars Briggs, Per Klit Christensen, Niels Damm, Finn Hansen, Martin Hesselsøe and Uffe Mikkelsen

Abbreviations and acronyms used in the texts or references

DNA	Deoxyribonucleic acid
EU	European Union
IUCN	International Union for the Conservation of Nature
LIFE	The European Union's financial instrument supporting environmental, nature conservation, and climate-action projects / L'Instrument Financier pour l'Environnement
NOVANA	Det Nationale Overvågningsprogram af VAndmiljøet og Naturen / National monitoring program of aquatic environments and nature.

I. Introduction

Denmark is a small country with a total area of 43,094 km². It consists of the Jylland (Jutland) peninsula, which is an extension of the European continent, and 443 named islands, of which 72 are inhabited. The largest islands are Sjælland (Zealand) and Fyb (Funen). The total length of the coastline is 7,314 km. The highest point is about 171 m above sea level, and the lowest one (behind a dike) about 4 m below sea level. The climate is influenced by the warm water from the North Atlantic Current. Average temperatures in January are about 0°C and those in July range from 16°C to 17.5°C from the coldest regions (northwest) to the warmest regions (southeast). Annual precipitation ranges from about 500 mm in the driest regions along the coast adjoining The Great Belt Storebælt (The Great Belt) to about 900 mm in the wettest region in south-central Jylland.

Fourteen species of amphibians occur naturally in Denmark. Although the regional differences in climate may seem small, they are large enough to influence the amphibian fauna. Some species are completely absent from Jylland and live only on the islands (*Bombina bombina*, *Bufotes viridis variabilis* and *Rana dalmatina*). Others live in East Jylland, but are absent or rare in the cooler parts of Jylland in the west, northwest and north (*Triturus cristatus* ad *Hyla arborea*). *Pelophylax ridibundus* lives only in Bornholm. *Rana temporaria* is absent from all southeastern islands. In 9 of the 14 species, the line demarcating the northern or northwestern border of distribution goes through Denmark. For instance, the absolute northern border of the range of *Ichthyosaura alpestris* is at 55°04' in southern Jutland.

Many of the localities for rare and threatened amphibians are on small islands and near the coast. All species living there tolerate at least moderately brackish water. Breeding success has been recorded up to the following salinities (parts per thousand), in some, but not all populations: *Lissoriton vulgaris* 3.5, *Triturus cristatus* 3, *Bombina bombina* 3, *Pelobates fuscus* 1, *Hyla arborea* 5.5, *Bufo bufo* 7, *Epidalea calamita* 7, *Bufotes viridis variabilis* 10, *Rana arvalis* 6, *Pelophylax esculentus* 5 (Fog, unpubl. observations). The adults tolerate even higher salinities and some species may be seen swimming at sea, especially in the inner Danish waters where the salinity of seawater is only about 10 parts per thousand. *Bufotes viridis variabilis* and *Pelophylax esculentus* may swim over distances of many km in the sea and may colonize remote islands in this way (Various unpublished direct observations and recordings of colonizations from the archipelago south of Fyn, Svinø Bugt, Øresund, Baltic Sea north of Bornholm; see also Hanström 1927).

II. Legal status of species

Since 1981, all species of amphibians are protected by the Danish Law of Nature Conservation. Collection of brood (eggs or tadpoles) of all but the rarest species is permitted, but collection of adult amphibians is allowed only for the commonest species, and then only for education or research. Table 67.1 gives the degree of protection of each species according to national law and to the EU Habitats Directive, as well as its IUCN classification in the Danish Red Data List.

III. Records of amphibians in Denmark

As early as 1847, eleven species of amphibians were known from Denmark (Lütken *ca.* 1847). A book on the Danish amphibians published in 1870 provided the first extensive list of localities for these species (Collin 1870). During the period *ca.* 1940–1955, there was much interest among amateur herpetologists to record and protect amphibians. Pfaff (1943) at the Zoological Museum of Copenhagen collected data on as many localities for each species as possible in order to produce precise maps of distributions. He dealt with the status of the rare species, and wrote of the firebellied toad: "There is then all possible reason to treat the last *Bombina* localities with the utmost care. Even if the disaster maybe cannot be avoided, it may possibly be postponed. It would of course be best to protect the animals and the localities totally." At that time, naturalists considered the main threat to amphibians to be the destruction of ponds, i.e., ponds were filled in, used as garbage dumps, or were drained. Especially in the period 1945–1955, naturalists were very active in protecting *Bombina*. They searched systematically for the remaining localities, negotiated with farmers to not destroy ponds, and sought to have ponds legally protected (unpublished notes and letters at the library of the Zoological Museum in Copenhagen). In most cases, they failed. However, in a number of places, they managed to convince the conservation tribunal that *Bombina* ponds should be declared as protected areas. The long-term effect of such protection has been very modest, however. Usually it merely delayed, not prevented, extirpation.

Table 67.1   Protection and classification of Danish amphibians.

Species	Protection under Danish law	EU Habitats Directive	Red Data List
Lissotriton vulgaris	Education and research	—	LC
Ichthyosaura alpestris	Strict	—	NT
Triturus cristatus	Brood only	II + IV	LC
Bombina bombina	Strict	II + IV	NT
Pelobates fuscus	Brood only	IV	DD
Hyla arborea	Strict	IV	LC
Bufo bufo	Education and research	—	LC
Epidalea calamita	Brood only	IV	LC
Bufotes viridis variabilis[1]	Brood only	IV	LC
Rana arvalis	Education and research	IV	LC
Rana dalmatina	Brood only	IV	LC .
Rana temporaria	Education and research	V	LC
Pelophylax esculentus[2]	Education and research	V	LC
Pelophylax ridibundus[2]	Brood only	V	DD
Special Pelophylax populations[2]	—	—	CR

Education and research: Adults may be collected when the purpose is for education or research. Brood may be collected freely.

Brood only: Adults may not be collected, only brood (eggs or tadpoles).

Strict: Total protection. Brood or adults may be collected only with specific permission from the Nature Agency of the ministry of the Environment.

CR = Critically Endangered; **DD** = Data Deficient; **NT** = Near Threatened; **LC** = Least Concerned.

[1] The form of the green toad that occurs in Denmark and in other regions adjoining the Baltic Sea is *variabilis* according to **DNA** analysis, i.e., variously interpreted either as the subspecies *Bufotes viridis variabilis* or the full species *Bufotes variabilis*.

[2] The water frogs in Denmark (*Pelophylax esculentus*) are not a species in the traditional sense because they originated as a hybrid between *Pelophylax ridibundus* and *Pelophylax lessonae*. In most countries, *P. esculentus* is a klepton, i.e., the frogs coexist with at least one of the parent species, from which it "steals" part of their **DNA** in order to persist. In Denmark, outside of Bornholm, *P. esculentus* is not a klepton, but lives only in pure populations, without either parent species. Their persistence depends on matings between diploid and triploid individuals. In Bornholm, the situation is different. There, *P. ridibundus* is present in some, but not most, populations of water frogs, and a few populations of pure *P. ridibundus* exist. Some of the mixed populations are peculiar, i.e., with one of the "species" occurring only as one sex. Most peculiar are those populations in which *esculentus* is present exclusively as males, whereas *ridibundus* is present exclusively as females. Such populations in which each "species" belongs to a different sex, are known from a few localities on Bornholm and the neighbouring island of Christiansø, and nowhere else in the world. These special populations are included in the Danish Red Data List and classified as critically endangered.

The greatest number of amphibian recordings were generated by Per Holm-Andersen (1927–1977), who had a tremendous impact on Danish field herpetology, although he never published anything and is not known by others than herpetologists (Hvid 2005). Per Holm-Andersen started to make precise field notes in 1942. He collected data mainly in the afternoon and night, and slept in a tent or in youth hostels. All observations, even of a common toad seen on the road, were entered on a map, and beginning about 1943 he started to keep a card file, in which all localities were precisely recorded. This file was kept up to date until 1975, and it contains 3,329 cards with information from Denmark, plus a few hundred from neighbouring countries. Some cards contain information from several localities, so altogether the file contains about 4,000 locality records. Most of the information is from 1942–1948. After 1948, he went abroad to study amphibians mainly in France, North Africa and Yugoslavia, but he was in Denmark again in 1958–1960, when he also compiled many records. In addition, he established a collection of animals preserved in alcohol.

The collection was placed in a cottage in Denmark, but the file, which by 1975 comprised about 2,700 file cards, was written in French, and although some of the animals were from Denmark, most of them were from the Mediterranean region. Shortly before Holm-Andersen died in 1977, he gave two young Danish herpetologists access to all his material, and that is how his data were passed on.

During the period *ca*. 1955–1975, amphibians in Denmark received scant attention, but in 1976 a thorough mapping project was launched. The initiative came from an organization of young field naturalists, 'Natur og Ungdom' (Nature and Youth). It was organized as a so-called 'atlas-investigation', in which the whole country was divided into squares of 5 km x 5 km, and people were invited to sign up for investigating one or more squares. The investigation ran from 1976 until 1986, and nearly 1,000 persons, of all ages and professions, contributed. Altogether, 1,052 squares, i.e., about half of the squares in the country, were investigated. Of these, 16% were investigated thoroughly, 37% to an acceptable degree, and all the rest only superficially. Thirty-six per cent of the squares also were visited at night to record calling amphibians. As a result, distribution maps were drawn for all species and were reproduced by Fog (1993).

A main result of the atlas investigation was that the last remaining localities for the rarest and most threatened species were pinpointed. It was clear that such species had undergone considerable decline since the survey published by Pfaff in 1943, and a close monitoring of how the remaining populations fared was requested. Amateurs and unpaid biologists made many crucial recordings. All former localities for *Bombina bombina* were investigated in 1976–1977 to locate the few remaining populations. A large systematic search during 1977–1978 surveyed all ponds with *Ichthyosaura alpestris* in Denmark (Bisgaard *et al.* 1979) and *Hyla arborea* was mapped in several regions.

The first official monitoring programme began in 1983 in the greater Copenhagen region (Northeast Sjælland), with mapping of localities for *Pelobates fuscus*, *Epidalea calamita*, *Bufotes viridis variabilis* and *Rana dalmatina*. The next initiative was a mapping of all ponds with calling *Hyla arborea* on the island of Bornholm in 1985–1986. At the start of the 1990s, most counties organized systematic mapping and monitoring of rare species, paying biologists to do the fieldwork. For *Hyla arborea*, all ponds with calling frogs in the whole country had been recorded by about 1993. Such professional monitoring is performed by a cooperating group of specialists, with a strong emphasis on recordings based on the animals calling at night, often supplemented by reporting on the breeding success (presence of tadpoles) in summer.

In most, but not all, Danish counties, monitoring of rare amphibians gradually grew into permanent yearly routines. This lasted until the end of 2006, when all counties were abolished due to administrative changes. From then on, some municipalities have more or less taken over the counties' monitoring, but most have not.

Most of the rare species are included in Annex IV of the EU Habitats Directive. This urged the Danish state to monitor these species. A regular monitoring programme was started in 2005, with surveys being repeated once every six years except for *Bombina bombina*, for which the interval is every two years. This programme is not all-encompassing, but rather consists of spot checks. Denmark is divided into 10 km x 10 km squares and, at each monitoring, relevant squares are randomly selected for investigation. Most selected squares are known to have the species, but some squares with no known occurrences are also checked, to be able to detect localities hitherto not known to be occupied. In each square, the investigator examines two pre-determined ponds plus two ponds of his own choice. So, when populations shift from one place to another, there is some flexibility in tracking such movements.

Up to now, there has been little use made of the data.

An internet portal, where amateurs can record all kinds of observations of plants and animals (www.Fugleognatur.dk), was launched in 2015 as a new 'atlas investigation' of amphibians and reptiles. Amateurs are encouraged to send in photographs of the animals they find and to have specialists identify the species. This 'investigation', however, merely systematizes all reported localities of these animals.

IV. The decline of amphibians

A. Rate of decline

Amphibians depend on bodies of water for reproduction, and if the number of such habitats declines, amphibians will also decline. To elucidate the rate of disappearance of bodies of water, many persons studied old maps and compared them with the present situation. They used maps dating back to about 1900 or about 1950, but in some cases as far back as 1868. Altogether, the fate of nearly 6,000 original ponds in many different places in the country was recorded in this way; in one county (South Jutland), a total count of all original ponds (17,000) existing in 1954 was made. The general conclusion from these studies is that the rate of disappearance ranged from about 2.5% to 25.0% per decade. A rough average for all areas and all sizes of ponds up to the late 1980s was about 10% per decade (equal to 65% per century).

The rate of disappearance of amphibians may be evaluated roughly from changes in distribution. For instance, we know that *Rana arvalis* has disappeared from nearly all of its former localities on Fyn (Funen). There also is much more precise information available on the rates of decline. Thus, in nearly all former localities of *Bombina bombina* after 1950, we know the exact ponds where the animals lived. When all such localities have been systematically inspected in more recent years, we can make a precise statement on a pond-to-pond basis on the rate of decline. In many cases we also know in which year the animals were heard for the last time.

On Bornholm, Arne Larsen, funded by a charity, began making systematic recordings of *Hyla arborea* in 1949. Recordings were marked on a map. Following his death in 1979 his widow kindly lent the map to one of us (K. Fog) in 1985 and 1986, to enable checking of the former localities. This enabled calculation of the rate of decline, and it was possible in nearly all cases to determine the approximate year in which the frogs disappeared.

More important than anything else, however, were the thousands of precise records made by Per Holm-Andersen (Fog 1988). These became available to the organizers of the 'atlas investigation' in 1977, just when this investigation had started. We got some funding to have all localities mapped in distinct atlas squares. Next, the file cards were photocopied and sent out to the respective investigators of these squares, with a letter explaining the purpose and asking the investigators to visit precisely the same localities.

During the course of the atlas investigation, many hundreds of former localities were investigated in this way. In addition, the administration of the greater Copenhagen region provided payment for professional re-investigation of localities in northeastern Sjælland – the region with the largest number of localities. Altogether, combining amateur and professional re-investigations, out of 3,329 file cards, 1,956 had been re-investigated, i.e. slightly more than half of them. About 100 of these were recordings of reptiles, which leaves more than 1,800 re-investigations of amphibians. Of these, about 1,500 involved breeding ponds and the rest were foraging localities. As Holm-Andersen often recorded more than one species per locality, the number of localities re-investigated is less than the total number of file cards. In total, about 850 breeding localities and 290 foraging localities have been re-investigated. This constitutes a unique dataset on the rate of decline of amphibians.

The main results of re-investigating Holm-Andersen's localities are as follows. The percentage of localities where amphibians have survived is about the same for foraging localities as for breeding localities. At foraging localities, amphibians were found again in 41% of the cases. At breeding localities, all species were still present at 35% of the sites, and some of the species still present on a further 8%, which means that at least one species was still present at 43%. After omitting doubtful cases, the places where the habitat was still present and not drastically changed, amphibians had on the average survived in 41% of the sites, whereas where the habitat was greatly disturbed or destroyed altogether, they had survived in only 1% of the cases. Thus, the result is that even when the breeding pond was still present and not greatly modified, amphibians nevertheless often disappeared. The rate of disappearance differs among species. Those species that were already rare declined by a much higher percentage than did commoner species. Data are given in Table 67.2. The table shows that the five rarest species survived in a much lower percentage of the cases than did the seven most common species.

Most of Holm-Andersen's observations were recorded in the period 1942–1948, but some were made in 1958–1960, and he also revisited a number of his own localities in 1968. From this it is possible to give some indication of the time-course of decline, and the result is that the greatest decline occurred after the 1960s. We have also other data on the time-course of decline for rare amphibians, especially in the case of *Hyla arborea* on Bornholm (see above). Furthermore, we have data on the decline of some amphibians, such as *Pelobates fuscus*, from the professional monitoring carried out during the 1980s.

When we combine all these data, and the data on the disappearance of ponds, we gain an overall picture of how the situation for amphibians evolved in Denmark from the 1940s until the late 1980s (Figure 67.1). In figure 67.1, the top set of curves (blue) gives the change in the number of ponds. The solid line gives the gross decline, i.e. the percentage of the original ponds still

Table 67.2 Amphibians' breeding localities in Denmark. Data recorded by Holm-Andersen mainly from 1942 to 1948 and revisited by participants in the atlas investigation from 1977 to 1986.

+ means positively recorded on revisit; (+) means most likely still present, but not actually recorded on revisit; ? means probably still present, probably not (undecided cases); (–) means most likely disappeared, but this is not absolutely certain; – means definitely disappeared (mostly cases where the pond itself has disappeared)

Species	+ and (+)	?	and (–)	Percent Survived
RARE SPECIES				
Bombina bombina	10	0	17	37
Pelobates fuscus	3	6	31	8
Hyla arborea	3	0	28	10
Epidalea calamita	6	8	11	24
Bufotes viridis variabilis	4	9	25	11
Rare Species Total	**26**	**23**	**112**	**16**
COMMON SPECIES				
Bufo bufo	95	14	59	57
Rana arvalis	144	54	124	45
Rana dalmatina	26	3	15	59
Rana temporaria	123	58	165	36
Pelophylax esculentus	164	48	184	41
Triturus cristatus	11	8	13	34
Lissotriton vulgaris	24	5	13	57
Common Species Total	**587**	**190**	**573**	**43**

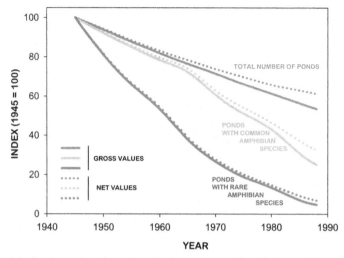

Fig. 67.1 Temporal decline in number of ponds and in the occupancy of ponds by amphibians in Denmark.

existing at any given time. The dotted line gives the net decline, which includes the effect of newly arisen ponds, e.g. ponds dug for hunting purposes.

The next set of curves (green) gives the decline for the common species, as listed in Table 67.2. At first, the decline is not much steeper than the decline in number of ponds but towards the end of the curve, the decline is much faster, indicating that in many cases the amphibians are extirpated even if the ponds are still there. There is also a considerable difference between gross decline and net decline – in the more recent years, quite a number of amphibian populations disappear from their original ponds, but move to other ponds and thus do not disappear completely.

The third set of curves (red) shows the decline of the rare species, as listed in Table 67.2. From the very beginning, the decline is much faster than explained by the disappearance of ponds. By the end of the period, in 1988, the rare species had survived in only about 3% of the ponds which they originally inhabited 43 years earlier. Most of the few surviving populations now have moved to other ponds, making the net decline not quite as large as the gross decline.

The rate of decline of these amphibians has been accelerating. During the first twenty years, the rate of decline for the rare species was about 32% per decade. From the 1960s until about 1980, it was about 44% per decade and, during the 1980s, it was about 67% per decade. So the amphibians were not just subjected to an exponential decline – they were subjected to a decline that was stronger than exponential. By 1988, the situation for amphibians, and especially for rare amphibians, obviously was severe.

B. Causes of decline

When a large number of old localities were revisited by experienced biologists in the 1980s, one could notice not only when a species had disappeared, but also why it had disappeared. When ponds had disappeared completely or were nearly destroyed, that was a sufficient explanation of the disappearance of amphibians. For ponds that were still there, one could see whether the degree of eutrophication was so great that the species were unable to breed. For ponds shaded by trees, one could see whether the shade was so extensive that the pond became too cold for breeding.

Figure 67.2 deals with the localities visited by Per Holm-Andersen. It shows the results for 307 localities in the greater Copenhagen region from which at least one species had disappeared. The second column from the left shows that for ponds that were completely lost, all species

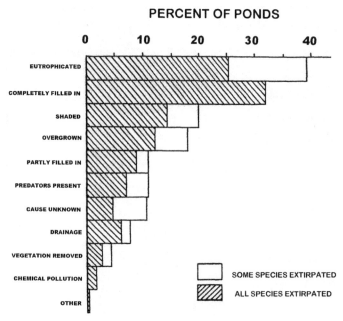

Fig. 67.2   Causes of decline of amphibians in 307 ponds in which at least one species had died out in North Sjæland (North Sealand), Denmark.

necessarily had disappeared. The same was true for the few ponds that had been polluted by chemicals, or subjected to miscellaneous other treatments (the two columns at the bottom of the graph). In the other categories of ponds that were still present, some ponds had lost some species but retained others (white parts of columns) but other ponds had lost all species (hatched part of columns). For all potential causes of decline, the taller the histogram, the more severe seem to be the effect of the agent of decline, especially the hatched part of the histogram. Thus, of the ponds that still remained, the most frequent cause of decline is eutrophication. There may be many causes of such eutrophication, the most frequent probably being drainage of water from fields, but also important are effluents from households, dumping of organic waste, and feeding of ducks. Total removal of the pond is only the second-most important cause of loss of amphibians. Shade from trees is a problem in nearly 20% of the cases. "Overgrown" refers to natural overgrowing ("pond senescence") that cannot be seen directly as a consequence of eutrophication. Predators are ducks or fishes. "Chemicals" refer only to situations in which barrels or bags with remains of pesticides were seen directly at the site; more diffuse pollution with pesticides could not be assessed.

Figure 67.3 deals with 80 localities where Arne Larsen recorded *Hyla arborea* on Bornholm, and where that species subsequently had disappeared. The categories on the X-axis are similar to those in Figure 67.2 but are separated into primary causes (hatched part of columns) and secondary causes (white part of columns). For instance, a pond may be heavily eutrophicated, and that condition considered to be the primary cause, but in addition, trees later grew so large that at present their shade would also be sufficient to eradicate the species; according to the owner, *Hyla* disappeared long ago, before shade became so extensive, so it is inferred that shade alone did not cause extirpation of the species. At another pond it could have been possible to have been heavily shaded early, with eutrophication subsequently being slow and gradual. In such a case, shade would be the primary cause, and eutrophication secondary.

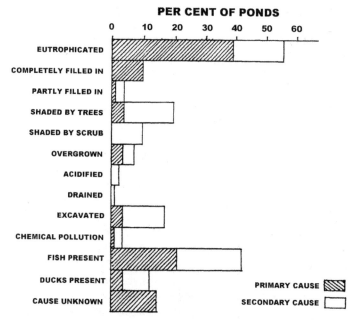

Fig. 67.3 Causes for the disappearance of treefrogs (*Hyla arborea*) from 80 ponds on Bornholm Island. For explanation of primary and secondary causes, see text.

We see that destruction of the pond (entirely or partially filled in) is the cause in only about 10% of the cases. Eutrophication is much more important, and is the primary cause in *ca*. 40% of the cases. Often it was possible to reconstruct the course of events directly. For instance, there may have been an outlet from the washing room at the farm via a ditch to the pond; the owners remembered when they bought their first washing machine, which would then have sent phosphate-containing detergents into the pond, and they remembered that the frogs then disappeared.

Hyla tadpoles are very sensitive to predation by fish. A typical scenario is that some person releases crucian carp into the pond, and from that date onward, *Hyla* did not breed there successfully. In every single case, this meant no recruitment of young frogs, and typically the last of the old frogs were dead after about seven years. This kind of event happened very often on Bornholm, and is considered to be the second-most important cause of decline there. The keeping of ducks acts in a different way. As soon as a pond is populated with ducks or ducklings, adult treefrogs leave the pond and stop calling there.

Figures 67.2 and 67.3 illustrate that the causes of amphibians' disappearance are most often not that the ponds disappear. Eutrophication; invasion by trees, bushes, or reeds; release or immigration of fish; and the keeping of ducks may all contribute to the gradual disappearance of amphibians in pond after pond.

The additive effect of all these causes has been massive. This was evident from the monitoring of rare amphibians in the 1980s. For instance, in the greater Copenhagen region, the localities known to be inhabited by *Pelobates fuscus* in 1983 were monitored in each of the following years. Out of 47 original localities, only 20 still contained this species five years later (1988), of which *Hyla* still bred successfully at 19 localities; five years still later (1993) there was successful breeding at only nine. Thus, the rate of decline was more than 50% in just five years! This decline was so impressive that it was obvious that merely passively monitoring made no sense. Monitoring had to be supplemented by active conservation measures.

V. Conservation measures

A. Protection of ponds

The first approach in Denmark to conservation of amphibians was based on the understanding that the main cause of declines was the destruction of ponds. Therefore, efforts were made to have specific ponds protected by declarations according to the Danish Law of Nature Conservation. Especially in the period 1945–1955, naturalists actively tried to benefit the species. They searched systematically for the remaining localities, negotiated with the farmers not to destroy the ponds, and sought to have ponds legally protected. In most cases, they failed. However, in some places, they did succeed. According to Danish law, a type of conservation tribunal has the final say if a declaration of protection is to be adopted, and in some cases the tribunal was convinced that this was a necessary measure. During the period 1946–1969, 22 localities inhabited by *Bombina bombina* were protected as single ponds, and 40 ponds containing *Bombina* were protected as a part of larger protected landscapes. Similarly, in 1981 and 1986, a number of ponds with their surrounding terrestrial habitat were considered important habitats for *Ichthyosaura alpestris* and were protected by declaration according to the Danish Law of Nature Conservation.

Such protection became redundant in 1992, when all ponds became automatically protected. This happened when a revision of the Law of Nature Conservation was passed in which it is stated that all ponds with an area of at least 100 m² are protected. The main reason for this amendment of the law was evidence from the atlas investigation demonstrating that all species of amphibians were in severe decline, and that the size limit of protected ponds should be as low as 100 m² in order for most populations to be saved. Thus, the information obtained in the atlas investigation was utilized politically to improve the conservation of these animals. The observance of this general protection of ponds was monitored by county administrations up to 2007 and by municipal administrations afterwards.

In 1981 and 1986, a number of ponds with surrounding habitats important for *Ichthyosaura alpestris* were likewise protected by declaration according to the Danish Law of Nature Conservation.

B. Mere protection of ponds is insufficient

We evaluated the effect of the specific legal protection of *Bombina*'s ponds mentioned above (Fog and Wederkinch 2016). As early as 1980, *Bombina* had disappeared from more than half of the protected ponds and, in the long term, it seems to survive in only 15% of protected ponds when no additional management is carried out. Protection may have delayed, but not prevented, extirpation.

That the mere persistence of a pond does not secure the persistence of its amphibian populations was a conclusion of a survey made by Fog (1997) who gathered information from Danish herpetologists on 780 ponds, of which 496 were unmanaged and 284 were managed in some way. The question was: what happens to populations of rare amphibians under the following scenarios: (*a*) the species recorded in the pond during a given year, and the same pond investigated at least five years later to see whether the species had survived; (*b*) the scenario the same as *a*, except that the pond was managed in some way, e.g. dredging, cutting of trees, or removal of fish. It was found that on average, the rare amphibian survived in only 40% of the ponds in scenario *a*, i.e. there was a risk of extirpation of 60% over a time span of just five, or slightly more, years. In scenario *b*, on the other hand, when the pond had received some active management, the rare amphibian had survived in 92% of the cases over the same time span. This pattern was found for *Bombina bombina*, *Pelobates fuscus*, *Hyla arborea*, *Epidalea calamita*, *Bufotes viridis variabilis* and *Rana dalmatina*. Thus, the conclusion was that the worst approach is to refrain from any management. The chance of saving a population is vastly greater when the pond is managed than when it is not.

It was noted in a previous section that when amphibians disappear from a pond, the main causes are (1) eutrophication; (2) shade from trees or bushes, and (3) introduction of fish. The active management required in these cases is to:

1. Remove the bottom-sludge where nutrients are stored, and block inflow of eutrophicated water.
2. Cut trees and remove willows, roots and all.
3. Pump all water out of the pond and let it dry out, preferably during July–September when the water level is lowest. In the case of crucian carp or goldfish, extreme care must be taken that all mud is completely dried out and remains so for several weeks, and/or all mud is dug out so that the bottom is completely dry, bare and hard, without the slightest amount of soft mud remaining.

Other potential measures involve removing drainage pipes or deepening the pond if it dries out too early, or raising the deepest parts of the bottom and making the banks more gently sloping if there is a need for the pond to dry out in dry summers (e.g. habitats for *Epidalea calamita*). Another option is to make entirely new ponds, by excavating a depression or by blocking drainage pipes and ditches. When an entirely new pond is created, or a formerly unsuitable pond is restored to make it suitable, rare species of amphibians will often colonize it. For 416 such cases, it was found that the ponds were colonized naturally in about 40% of the cases, and equally often if the pond were new or restored (Fog 1997).

C. Initiatives on a national level

The Danish government launched a new initiative, nature restoration, with extensive funding from 1986 onward. This supplied funds for carrying out extensive projects in many counties. Most of the projects reviewed in the following paragraphs have been financed by county administrations or other public authorities. Some other projects have been funded by charities, and some have been initiated and/or financed by idealistic private persons.

In connection with recent reorganization of the ministries, we now have "Naturerhvervssty-relsen", an agency for agriculture, forestry and fisheries. This agency launched a type of financial support for projects aiming at animal species on Annex IV of the EU Habitats Directive. This action includes the dredging and digging of ponds and restoration of stone walls. Private landowners, as well as organizations, consulting firms and municipalities can apply. Money will begin to be spent in 2016. Many farmers and other landowners are expected to apply. It remains to be seen whether this will be an effective process. A main concern is whether the new ponds will be constructed close to small, threatened populations.

VI. Species of special concern

In the following paragraphs, we report in detail on the efforts to protect the rarest and most threatened species, with an emphasis on active management. This is not a full list of all projects as some also have been carried out for the commoner species, such as *Triturus cristatus*, *Rana dalmatina*, and *Rana arvalis*.

A. Alpine newt, *Ichthyosaura alpestris*

The alpine newt was found for the first time in Denmark in 1949. It occurs only in southernmost Jutland, in the woods and forests west and southwest of Aabenraa and in woods at Flensborg Fjord, close to the Danish/German border. It is one of our rarest species. In 1975 a group of five persons formed a task force with the purpose of mapping its distribution in Denmark. Searches for larvae by using a large dipnet were carried out during the months August–September. Out of 600 ponds investigated, mainly in 1978, the species was found in 46 of them. Most of these were small forest ponds less than 100 m². As stated above, some of the ponds became legally protected in 1981 and 1986.

The next bout of searching was carried out in 1988. By then, the number of occupied ponds had dropped to 32. The task force received some funding to manage ponds, and other herpetologists were consulted about how to use the money most wisely. It was decided to focus on peripheral populations in the most precarious situations because disappearance of such populations would mean a contraction of the entire range.

Management was mainly dredging and removing fallen leaves, but also in some cases new ponds were made. This was very successful. From 1988 onwards, all known populations survived, even the population in a wood where no breeding pond still existed, but where a few adults were still present on the forest floor.

The status in 1997 (Bringsøe and Mikkelsen 1997) was that 76 ponds had been restored/dredged, 14 former ponds had been re-created, and 121 new ponds had been made. In response, the number of ponds naturally colonized by the species increased rapidly. More than 100 ponds were inhabited by 1992 and 201 were inhabited by 1997. Seventy-five percent of all new and restored ponds had been colonized. Since then (up to 2009), a few more ponds have been made and some of the old ponds are still being dredged at intervals. The distribution has expanded slightly to the north and to the west, but a barrier (an industrial area in the northern part of Aabenraa) has prevented the species from colonizing the forest northeast of that city.

In summary, the efforts of active management, with focus on the most endangered populations first, have been extremely effective, with the number of localities multiplying by six-fold in just nine years.

B. Fire-bellied toad, *Bombina bombina*

The fire-bellied toad has always been relatively rare in Denmark. In 1970, about 23 populations survived, with occurrence in a total of about 100 ponds in the whole country. By 1988, this had declined to seven populations with occurrence in only 28 ponds, of which only 10 ponds showed breeding success. The total number of individuals was about 750–800, as calculated from capture-recapture studies at all localities.

It was a high priority to stop this drastic decline, and starting from 1982, much effort was made to save the few remaining populations. For the most threatened populations this involved artificial breeding. The first artificial breeding occurred in 1982–1983 on one of the small islands south of Fyn, where eggs were collected from a breeding pond and reared in captivity. The resulting toadlets were released in newly dug ponds a few kilometres from the original site. In 1984, this was followed up by the construction of 'breeding cages' with the bottoms and sides all about 2 m². Such cages were placed in shallow water, and adults of both sexes were placed there so as to mate and spawn inside the cage. After spawning, the adults were returned to the pond, and the eggs were left to develop within the cage. Predators of all kinds were removed to increase the survival rate of the tadpoles. This procedure was successful, and soon the first prototypes of breeding cages were replaced in greater numbers by professionally made cages.

Unfortunately, the adults were not always willing to spawn in the cages. In one population on Fyn, only three males could be found in 1984; two females from another population 7 km away were imported and placed in cages with two males; but no spawning occurred. The following year, the population had disappeared.

Things were about to go the same way in the last population on Fyn, where the last breeding pond suffered from the release of fish. The first attempt in 1982 to eradicate the fish was not completely effective. From 1984 onwards, the few surviving fish had again multiplied and prevented breeding by *Bombina*. The nature authorities did not allow poisoning of the fish with rotenone. In 1987 we attempted instead to place adults in a cage, but all attempts to have them breed either within or outside of the cage failed. This was repeated in 1988, with the additional feature that now both males and females were injected with a sex hormone (chorion gonadotropin). Still, there was no spawning. Two very new ponds, however, were made close to the old pond, just separated from it by a narrow wall. These small ponds were fish free, and freely roaming toads entered them and spawned there. When the breeding cage was moved to one of the small fish-free ponds, the toads in the cage mated and spawned immediately. So we learned that spawning occurs only if the toads cannot smell the presence of fish in the water.

On the island of Ærø south of Fyn, only one single male remained in 1986. No females were left. The male was collected and allowed to hibernate in captivity, and the next spring it was put in a cage with some females on another island. It was injected with sex hormone, and spawning occurred. The offspring were reared and returned to the original locality where, in the meantime, the pond there had been dredged.

On Enø, a small island near Sjælland, *Bombina* had survived in three mutually isolated ponds. The total population was about 19 animals in 1988, and 9 animals in 1989. The population could be saved only by artificial rearing. In 1988, males and females were brought together in breeding cages placed in their own ponds; when they entered amplexus, the amplectant pair was moved to a 25 l aquarium containing water plants and heated water, and spawning occurred overnight. This method was necessary partially because water quality in the original pond was not good enough.

In 1989, efforts were made to mate males from ponds that had a male surplus with females from ponds containing a female surplus. This was not easy. When males were moved to cages in other ponds, they stopped calling and would not mate. So the procedure was to keep the animals in cages in their own ponds. Then, when a female was ready for mating, it was transported in a bucket to that pond and the male, while calling, was captured and immediately placed in the bucket with the female. This always resulted in mating. The amplectant pair was then placed in a heated aquarium. Some animals were not in a mood for breeding and had to be treated with sex hormones before mating was possible.

The result on this island was a viable population with more than 100 individuals. Unfortunately, after 2000, some negative trends appeared and very few eggs were laid, for no obvious reason. Some new negative environmental factors emerged; the population suddenly crashed, and the last few individuals were brought to captivity. Their offspring now all live in captivity. Some offspring have been released back into nature, but have not survived, most likely owing to predation by herons.

The latest new measure taken for this population is to have an enclosure on a piece of land owned by the municipality. The enclosure is 2,500 m², with two newly dug ponds, each of 250 m², surrounded by a fence that prevents toads from emigrating, and covered by a net to prevent access by herons and gulls. Some captive animals have been placed there and will hopefully produce a 'stock' population, from which surplus offspring can be taken for release in suitable natural ponds in nature.

Another small island south of Fyn had *Bombina* in a single cattle-watering pond. The pond was flooded by sea water during a winter storm early in 1976. Luckily, the owner pumped it dry to remove the salt water, and *Bombina* bred successfully afterwards. It had a stable population of about 50 animals. Then, again the island was partially flooded by salt water in the winter of 1995/1996, and presumably many individuals died when their hibernation site was flooded. The population never recovered from these conditions, and in 1998 only four adults were left. They were placed in two rearing cages and injected with hormones. This resulted in spawning and the offspring were reared in captivity. The rearing continued for three generations in captivity; the first captive generation was very badly hit by inbreeding depression, and most animals died, but for each new captive generation, the health improved. Offspring have now been released back onto the island, and the wild population there is now larger than ever before.

All such efforts with artificial breeding were combined with the restoration of ponds and digging of new ponds, so as to have places for the reared offspring. Also the larger populations elsewhere in the country, where no artificial rearing was necessary, were benefited by projects of various kinds. This consolidated the populations and several new ponds were colonized. In some cases, ponds were improved just "at the last minute". In the next spring, the surviving one or two males called very loudly and persistently in the night, thereby attracting toads from other ponds. Capture-recapture evidence shows that males and females were attracted from other ponds 1–2 km away. In this way, the restored ponds were quickly repopulated. One pond had been pumped dry to eradicate fish, with success. After ten years, the toad population there had grown from extremely few to about 500.

A national survey in 1996 showed that until then, 89 projects had been carried out for *Bombina bombina*. In 62 of these there were no *Bombina* in the pond before the project. In 40 of these, reared offspring were released with success and in 19, *Bombina* colonized naturally.

The year 1999 started a new phase when an EU LIFE project, "Consolidation of *Bombina bombina* in Denmark" was initiated. This meant that a further number of ponds could be dug or restored. In addition, it marked the realization of a new concept, 'mirror populations' or 'reserve populations'. For the four largest populations, offspring were reared in large numbers and released at completely new sites. At least 2,000 newly metamorphosed toadlets were released at each new site. For instance, offspring from the island of Nekselø were released at the peninsula of Røsnæs, 20 km away. The aim was to establish a new population with precisely the same genetic composition as the original one. So if some time in the future the original population at Nekselø becomes threatened, it will be possible to add supplementary animals to the population, or renew the population without a loss of genetic variation if it died out completely. Care was taken to obtain offspring from as many parent animals as possible, in order to transfer nearly 100% of the genetic variation of each parent population. In connection with the project, samples of DNA were taken and analyzed at a German university; in all cases nearly 100% of all alleles present in the original population were retained in the newly established mirror populations, which thus, really were 'mirrors' of the original genetic constitution.

Similar transfers were made from three other populations, where such transfers had already begun a few years before the start of the LIFE project, e.g. to populate corridors in the landscape. In these three additional cases, the LIFE project also meant the digging of many new ponds and the release of reared offspring at new sites.

In one locality, Enø, to which reference was made above, the LIFE project came too late. No suitable new site could be found in 1999, and establishing 'mirror populations' such as that on Enø began only in 2004, with the start of yet one more LIFE project. By then, however, the original population had started to decline severely and, although more than 800 reared toadlets were released at the new site, none survived. The end result was that the new population failed to become established and the original population died out in nature and only survives as a captive population.

On a national scale, the entire situation for *Bombina* has improved. Hundreds of new ponds have been made, and the total Danish population has increased from about 800 in 1988 to 1,700 in 1998 and about 2,000 at the end of the first LIFE project in 2003.

However, in most cases progress has not continued. As long as many active measures are taken, and animals are reared and released, the populations increase, but when the effort ceases and the animals are left to make good with what they have, moderate or severe backlash occurred in several places. There are interesting differences between the populations. One of the populations (from Nekselø) seems to be especially strong and fit. The reared offspring released at the new site, Røsnæs, has survived extremely well; they have bred successfully there, have colonized new ponds farther away, and have appeared up to six km from the ponds of origin, even though they have had to pass through less favourable landscapes. Similar strongly positive trends have not been observed in the other populations.

To explain the different outcomes in different populations, we have tried to construct a scoring system for negative factors. The score is composed of the following elements:

1. Number of ponds with breeding success
2. Total water surface of ponds utilized for feeding
3. Distance from ponds to terrestrial habitat used in late summer
4. Distance from ponds to the most likely hibernation sites
5. Percent of surrounding landscape cultivated
6. Size of genetic bottleneck of the founding population during the past 20 years
7. Frequency of herons hunting at the ponds
8. Age of offspring when released (tadpoles; newly metamorphosed animals; adults)
9. Number of offspring released per pond

The combined score shows some correlation with the fate of each newly established population and the trend in existing populations.

C. Garlic toad, *Pelobates fuscus*

The garlic toad (or spadefoot toad) is often difficult to find, and potentially many Danish populations may not have been recorded. However, in the recent monitoring program NOVANA, organized by the state, squares are selected randomly, and suitable ponds are investigated within them. In these randomly selected suitable ponds, populations are rarely found. Only in one or two cases have populations been found that were previously entirely unknown, and only in few cases have garlic toads been recorded at distances of a few km from their nearest known occurrences. This indicates that most, but not all, populations in the country have been found and recorded.

The situation for the garlic toad is very different between east and west Denmark. In east Denmark this species occurs on Sjælland and some of the islands south of Sjælland, but it is rare and has declined severely. It is extinct in most parts of this region.

The largest populations in east Denmark are in north Sjælland. Systematic monitoring started here in 1983, and by 1988 it had become clear that active management of ponds was crucial for the preservation of the species.

We have elaborate statistics for garlic toad ponds in the former Frederiksborg County for the years from 1983 to 2005. Garlic toads were recorded at least once in each of 64 ponds. Thirty-three ponds have been improved in some way, e.g., dug as new ponds, dredged or excavated, or managed by removing willow scrub. In unmanaged ponds, average numbers of toads were much smaller by the end of the study period than at the start. In managed ponds with stable occurrence of the toads, the average number had increased somewhat, and in managed ponds with only sporadic occurrence, the average had also increased. Altogether, out of 64 ponds, the species had survived in 21. Only in 13 ponds was there a positive trend. The net trend for all ponds was a decline.

After 2005, renewed efforts have been made to increase the number of suitable ponds in the area, partially financed by EU Life projects. Up to now (2016), about 50 ponds have been dug or improved, and ponds that were overgrown by reeds (*Typha, Phragmites*) or shrubs (*Salix cinerea*) or had their surface covered by duckweed, *Potamogeton,* or other kinds of floating plants, have been managed repeatedly. As a result, the overall negative trend seems to have stopped, and there are some cases in which populations have expanded and colonized new ponds.

In the rest of Sjælland, the situation is much worse. There, garlic toads have been found mostly in single, isolated ponds, and nearly all efforts to save such populations have failed. Even where the present conditions are now excellent – e.g. with newly dredged ponds – the populations gradually disappear. They are probably too inbred to survive and/or the efforts to improve the ponds should have been supplemented by intensive artificial breeding.

There is one single case in north Sjælland where an isolated small relict population of garlic toads was saved. Garlic toads were recorded in low numbers in two or three ponds, none of which were ideal for the species. One pond, on public ground, was dredged in 2010. In 2011 it was fenced to capture any toads migrating to the pond, but only two males were caught. In another temporary, desiccating, wet depression in the middle of a field, ½ km from there, a few males were heard, and a string of eggs was found. These eggs were reared, and the offspring released into the dredged pond. It seems that a viable population has resulted.

On the island of Lolland, garlic toads were recorded in several ponds in 1982. In 1992, only one male could be heard in one single isolated pond, which, however, had been partially filled up and was too shallow. It was made slightly deeper already in the same autumn, and a small population of toads managed to survive and increase in number until 2004, after which it disappeared quickly. Inbreeding may have played a role. However, a new large metapopulation was discovered in 2006 elsewhere on the island. The species was recorded in 11 ponds there, with large populations in some of them. Five new ponds have recently had their connectivity with other ponds increased.

In west Denmark, the garlic toad used to be widely distributed in Jutland and a few adjoining islands. It has disappeared in most regions dominated by clayey soils, but has survived in many smaller or larger groups of ponds on sandier soils.

In east Jutland, it has survived relatively well on the largest peninsula, Djursland. There, a large project on ponds was started in about 1990. Aarhus county announced that people could apply for having ponds dug or dredged with the county paying 25–50% of the costs. A machine began to dig at one end of the region and excavated the ponds in systematic order, finishing at the other end several years later. At least 400 ponds were made in this way and, of these, about 20 new ponds were placed where they would be of benefit to garlic toads. By 1996, three of these had been colonized. One further pond had been colonized by 2001. However, several new ponds within 200 m of existing breeding ponds had not been colonized.

At Djursland, out of the 41 ponds occupied by garlic toads in 1990/1991, they had disappeared from 23 by 2001, i.e. a gross decline of 56%. There was only one recent colonization, indicating a net decline of 54%. These data illustrate that making a lot of ponds over large areas to the benefit of the general biodiversity provides little assistance to specific, threatened populations of rare species.

In 2005, amphibian specialists were contacted when a large golf course was being constructed near the core area for garlic toads on Djursland. At the start, there were about 20 ponds in the area, and garlic toads occurred in 8–12 of these. Much care was taken to protect existing localities e.g. by buffer zones, and to define habitats close to the pond where nature had higher priority than golf. About 20 new ponds were made. The results were positive. In 2015, ponds with garlic toads had increased in number from *ca.* 10 to 17, a 70% increase. However, garlic toads were not as responsive as other protected amphibians; ponds with *Triturus cristatus* and *Rana arvalis* increased in 30 or 31 ponds.

Farther south in Jutland, initiatives for the garlic toad began in Vejle county in 1996. From about 2000, some ponds were made in an EU LIFE project for the great crested newt, and some of these were also of benefit to the garlic toad. Yet another LIFE project ran in 2004–2006, this time focusing directly on the garlic toad.

Altogether, up to now, attempts have been made to save 13 mutually isolated populations of garlic toads in this region, the former Vejle county. A total of 67 new ponds have been dug, and 18 old ponds have been dredged specifically for this species. Out of the 13 populations, 3 to 5 have disappeared in spite of this effort. This may be because of immigration of fish, or it may be because of inbreeding. Five populations have been rescued just in time, owing to dredging of their last breeding pond. The remaining three populations probably would have survived in any case, but have now been consolidated and have spread to neighbouring new ponds. New ponds have been colonized only in the two largest and least-threatened populations. In the small, barely surviving populations, none of the newly dug ponds have been colonized up to now.

Artificial breeding and release from collected eggs have been performed in the years 2008 to 2012. Offspring from one population have been released in another population, and *vice versa*, to counter any inbreeding. They were released as large tadpoles or newly metamorphosed toadlets.

In an area characterized by hilly end-moraines, seven or eight ponds were dug after 2000 for the benefit of *Triturus cristatus*. In 2011–2012 these also were used to accommodate a reserve population of garlic toads. Eggs were collected and reared from three populations in the region, and released into the new site. The result was excellent. In 2014, garlic toads were calling in all seven ponds.

In the former county of South Jutland, 16 ponds were made for the garlic toad up to 1996. In six cases, this consisted of the dredging of existing breeding ponds to secure the survival of the species. Out of three new or restored ponds where the species did not live before, one has been colonized.

On the island of Als close to south Jutland, little has been done for the garlic toad, and by now the species has survived in very few ponds. Some nature projects close to these, e.g. with improved grazing, may probably become of some help.

The largest populations of garlic toads in Denmark were found in the utmost southwest of Jutland, at Hjerpsted, in 1992. There, out of 76 ponds investigated in the core area, calling males were heard in more than half, and tadpoles were found in 28%. The populations were in a phase of severe decline, with tadpoles in only 17% of the ponds four years later. By then a large pond project was carried out from 1996 to 2000. There, and in neighbouring areas, 146 ponds were dredged, and 36 new ponds were made. The toads responded positively, and at the next census in 2000, the trend had been reversed. By then, tadpoles were found in 31% of the investigated ponds, including some of the new ponds. All of this was financed by the county. In the administrative reform of 2007, the county was abolished, and since then all efforts there have ceased. As a result, the ponds have again deteriorated, and populations have declined once again. Most of the ponds are small ponds in cultivated fields, with little suitable terrestrial habitat outside of the fields, and the modern scaling-up of agriculture (larger fields, larger machines) has contributed to the decline.

Another area where the species was supported by pond projects in the 1990s was between Aabenraa and Haderslev. There again, all support ceased when the county was abolished, and the recent situation is that the status of the species is back at the level prior to the pond projects.

A more lasting success has been obtained in the southeastern part of south Jutland, where a new motorway has been built. Garlic toads occurred in the area, and this released money for replacement ponds. This promoted progress for the species, which now has stable populations in more than 10 ponds there, owing not only to the ponds themselves, but also to appropriate improvement of nearby terrestrial habitats, such as raising the water level of moorlands.

In the former Ribe county (southwestern Jutland), nine breeding ponds for this species had been restored by 1996, mainly in the sandy regions in central Jutland. The species had survived in six of these. In addition, 39 additional ponds were made or restored. Nine of these have been colonized. The total effect of these 48 ponds up to 1996 was a net improvement in prospects for the species.

After the abolition of the counties, no monitoring was made there until about 2011. By then, some smaller populations had died out, but large core populations had survived. A few pond projects were then carried out by the municipality, but without assistance of experts on amphibians, it bodes a relatively low chance of success.

For many years, no populations were known in west-central Jutland (the former Ringkøbing County). However, two populations of the species were found there about 2000, and these have been supported by pond projects from 2001 onwards. This has led to expansion of the populations and colonization of new ponds, but also some backlashes, e.g. due to release of fish into some ponds.

In the northern parts of Jutland (the former counties Viborg and Nordjylland), the garlic toad has become widespread in some regions, but has strongly declined everywhere else. Up to 1996, only ten ponds were made for the species there, and a few of these have actually become colonized. In recent years, fewer than ten additional ponds have been made. At least three have been colonized.

In the region around Viborg, garlic toads were recorded in many ponds up to 1992. No measures to preserve the species have been taken there. Out of the ponds where the species occurred in 1992, 63 were investigated again in 2009. The species had disappeared from about 75% (in just 17 years). A main cause of the decline was the release of fish into many of the larger ponds, but still nothing is being done for the species in this region.

There are some interesting cases in the northernmost part of Jutland (Vendsyssel). One case is a pond on a property with organic farming and with emphasis on the cultivation of onions. This pond has probably the highest population anywhere in Denmark (about 300 calling males), so it is possible to have large populations if the surroundings are optimal (also in Estonia, onion cultivation is known to be favourable for this species). It occurs as an isolated population, but the prospects are good, as organic farming is expected to expand, with creation of several additional ponds.

In a dune area in northwestern Vendsyssel, 1–2 km from the coast, is a small surviving population. In 2015, eggs were collected there and reared, and the offspring were released back into the original site plus into a new site. The success of this project cannot yet be evaluated.

To sum up the situation for Jutland, there are very mixed trends. Where pond projects were financed by the counties, the initial positive effects have been lost again, especially when the ponds are among fields and there are no suitable terrestrial habitats. On the other hand, where the projects have been monitored closely by herpetologists, and where there have also been opportunities to create or preserve adjacent terrestrial habitats, there have been large, and probably more lasting, successes.

D. European treefrog, *Hyla arborea*

The treefrog used to be distributed over the whole of Denmark except for middle, western and northern Jutland. However, already by 1900 it had already disappeared from several regions, and the decline has accelerated since then. This species disappeared from north Sjælland about 1950, from central Sjælland about 1960, from Møn about 1950, from Falster about 1960 and from Fyn (Funen) about 1975. On a national scale, it disappeared from about 90% of its former breeding ponds from the 1940s to about 1980 (Fog 1988), rising to a loss of 97% by 1990.

The first efforts to counter this negative trend were made on Bornholm in 1983, when 14 treefrog ponds were dredged. In 1985–1986 the distribution of treefrogs was mapped on the whole island, with payment from the county. Calling males were heard in 162 ponds, which was a marked decline from the estimated 1,000 ponds in about 1950. The decline was obviously accelerating, with a 70% loss in number of ponds with calling frogs from 1976 to 1986. An advertisement in a Bornholm newspaper calling for applications to have ponds dredged with money from the county resulted in applications from *ca.* 150 farmers. All applicants were visited in order to find out what projects would have highest priority. From 1985 onwards, large numbers of ponds were dredged and even more new ponds were dug. Up to 2001, 770 ponds were made available, of which 426 were new ponds (Hansen 2004). Since then, few additional ponds have been made.

The results have been very positive (see Figure 67.4). Although the number of ponds with calling males still declined up to 1991, reaching a low of 70 ponds, it has since then increased very much, with about 600 ponds being colonized naturally since 1991, i.e. the number of occupied ponds has increased ten-fold. The number of calling males has increased roughly in proportion to this and the distribution has expanded.

The situation is somewhat similar on Lolland. There a mapping project in 1981–1982 revealed 109 ponds with calling treefrogs. The next mapping in 1991 showed a drop to 32 ponds, i.e. a 70% decline in ten years, as on Bornholm (treefrogs were also found in a few additional ponds not investigated ten years earlier).

In the county where Lolland is situated, a local nature-restoration project was formulated in 1991 in which the total budget included first a mapping of the remaining populations – with the

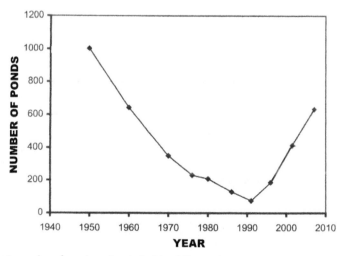

Fig. 67.4 Change in number of ponds on Bornholm Island, Denmark.

Data before 1976 are calculated from knowledge of the rate of disappearance during each decade. Data from 1976 and 1981 are from Arne Larsen and Finn Hansen, data from 1986 are from Kåre Fog, and data from 1991 onwards are from recordings made by Finn Hansen.

results indicated above – followed immediately by restoration of ponds at relevant sites found during the mapping.

During the period August 1991 to February 1992, 30 pond projects were carried out. Further pond projects were conducted since then. During the whole period 1991 to 2006, 137 pond projects aimed at treefrogs were carried out. Of 101 ponds that were dredged, in 11 of them fish were eradicated; also 19 new ponds were dug. The overall effect was that the total population of treefrogs increased from about 250 males calling in 39 ponds in 1991 to 1,100–1,400 males calling in 133 ponds in 2001–2002.

The original larger groups of treefrog ponds, concentrated geographically in a few clusters, by 1991 had become fragmented into several smaller, mutually isolated clusters or single ponds, designated in Table 67.3 by A to R. Based on several lines of evidence, it can be assumed that there was no contact between populations situated more than 4 km apart. This assumption was corroborated by analyses of DNA (Andersen *et al.* 2004; Fog and Andersen 2005).

Table 67.3 Survey of clusters ("populations") of treefrogs (*Hyla arborea*) at ponds on Lolland and Zealand. In each row, the narrowness of the population bottleneck is indicated as the number of males at the low point of the first population decline after 1988. The year of this low point is indicated. Note that populations A and B are called S and T by Andersen *et al.* (2004). n.i. = not investigated.

Population	Year	Low Point (number of males)	Percent of Eggs Reaching Metamorphosis	Development of Population up to 2014
LOLLAND				
A	1991	8	59.5	Admixture of genetic material from population B in 2003. Population size *status quo*
B	1992	6	42.0	Admixture of genetic material from population A in 2003; extirpated 2008
C	1991	100	81.7	Strong increase
D	1992	2	n.i.	Probably connected to C and E. *Status quo* explained by steady immigration (rescue effect)
E	1992	3	70.0	Increase
F	1991	25	72.1	Increase
G	1993	2	n.i.	Probably extirpated 2005. Reintroduction from F and H in 2007–2008
H	1997	3	43.1	*Status quo*
J	1991	2	81.4	Strong increase
K	1993	6	54.9	In 1998–2000 again few (0–3), later about 10
L	1995	4	56.1	Increase
M	1991	3	82.4	Increase
N	1993	1	n.i.	Extirpated 1994
P	1992	11	79.7	Strong increase
Q	1992	7	65.2	After first recovery, again a decline to near extirpation; now rescued by improved contact to P
R	1992	4	n.i.	Extirpated 1994
SJÆLLAND				
Stensby	1989	1	65.1	Strong increase and expansion
Præstø	1989	5	49.2 (1990) 75.0 (1990s) 56.9 (2005–2006)	Strong increase; little expansion; decreased again
Rejnstrup	2003	22	71.4	Slight increase and expansion
Slagelse (mixed population)	Started 1994		86.0	Strong increase; strong expansion

Some of the isolated populations on Lolland were extremely small when they were rescued by pond projects. The third column in Table 67.3 shows how few males were calling when the populations were at their minimum. Most of these extremely small populations have survived up to now, but one might expect that they would be affected by genetic erosion ("inbreeding depression"). To monitor this, eggs were sampled in nearly all populations and brought to the laboratory, where they were reared until just before metamorphosis. The result was that the fitness (percentage of eggs viable up to metamorphosis) was greatest in the largest populations and significantly smaller in most of those populations that had passed through narrow bottlenecks (Andersen *et al.* 2004, and Table 67.3). In addition, those populations that showed a reduced fitness (survival) in rearing experiments also fared relatively badly in nature. Their population size grew little, and they had very low ability to colonize new ponds when such ponds were available. That is, although it was possible to save several populations with extremely low size (2–8 males), most of these had suffered permanent genetic damage, indicating that inbreeding is a problem to be taken seriously.

On Sjælland, only three original populations survived (Stensby, Præstø and Rejnstrup). The first two of these were extremely small when the rescue effort began and, although they narrowly escaped extirpation, they managed to recover to approximately their original size. As shown in Table 67.3, these too had a markedly reduced fitness, with maximally 75% of the eggs being viable.

In the two smallest Sjælland populations, the rescue effort was a combination of the restoration of ponds (e.g., dredging, eradicating fish, digging of new ponds) and artificial rearing. Eggs were collected in the ponds, reared, and the offspring released either at the same site or at other sites a few km away. Up to 7% of the released froglets survived to maturity, and the populations became well established. As a follow-up, more pond projects were carried out. At Stensved, 48 ponds were constructed up to 2009, and a further *ca.* 60 ponds since then. About 30 ponds are now inhabited, and the frogs have expanded up to 5 km from their origin.

Thus, at Stensved there was a very small population, with only four calling males in 1988 when rearing began, and only one male left in 1989. In spite of the poor odds, it has survived. Growth and expansion was slow for the first *ca.* 10 years, but then proliferation started, and now it is a large, widespread population. Thus, saving of the smallest population possible has occurred.

For the other two original populations on Sjælland, success has been much lower, and up to now only a few new ponds have been dug in their vicinity.

The absolute northern limit of distribution of *Hyla arborea* is in Jutland. The limit used to be at about 56°20′, but the species has survived only up to 55°40′, just south of the city of Vejle. The number of ponds with calling treefrogs south of Vejle reached a low point of 11 ponds in 1988. Vejle county paid a herpetologist to contact landowners and initiate pond projects and, in less than ten years, more than 100 ponds had been made, and the population expanded greatly. The expansion is still ongoing now (2016).

Farther south, south of Kolding Fjord, treefrogs survived in eight ponds, and a similar project there gave similar results. In the county of South Jutland, treefrogs had survived in about 75 ponds on the mainland when pond projects started in 1991. The species was widespread in the vicinity of Haderslev, but there was a local limit to distribution farther south, halfway to Aabenraa. Near this limit many ponds were made, with considerable success, including the creation of a new large population that served as a source. From there, frogs have colonized additional ponds farther south, at distances up to 5 km, where 20 ponds have been created from 2003 onwards, mainly through private financing.

In all these populations in Jutland, treefrogs occurred in larger clusters of ponds, not in single isolated ponds. The one exception was at Rødding in southwestern Jutland, where treefrogs called in a single, completely isolated, cattle-watering pond. Although a number of additional ponds were made, and although rearing and release was attempted, the effort to save this population failed. It was the only population in western Jutland, and this is the only case where attempts to save this species have failed.

The island of Als also belonged to South Jutland county. The populations there were mapped on an amateur basis in 1985–1987: treefrogs were still widespread, but with few ponds remaining (*ca*. 20 ponds with calling males). An intensive effort to dig and improve ponds in about 1990 reversed the downward trend, and the total population increased very much. For over 20 years, no further measures were taken; in favourable habitats, the increased populations have remained high, but in the general agricultural landscape, a new decline is gradually becoming apparent.

An overview of the total effort to save treefrog populations in Denmark was published by Fog (1997). Data had been gathered for all projects up to 1996. By then, a total of 1,177 ponds had been made or restored for treefrogs. The results on a national scale were very positive. The total number of calling males increased from about 3,700 to about 9,000 from 1991 to 1996. Thus, it is quite rewarding to work on conservation of treefrogs – there is usually a quick, positive response. In addition to saving authentic surviving populations, a number of new populations have been established by the release of captively reared offspring.

The first major project was to bring treefrogs back to the natural areas just south of Aarhus in east Jutland, where they had lived until the 1950s. The theory to be tested by this project was that treefrogs are fully able to thrive near their former northern border of distribution, and that declines are not due to climatic changes. The project was financed by the municipality and started in 1982, when a number of ponds were created; by 1988 more than 100 ponds had been made (Skriver 1988). During the years 1985–1987, more than 6,000 treefrogs were reared from eggs collected from the Vejle population. The first calling males were heard at Aarhus in 1986 (three males) and 1987 (10 males). Since then, more ponds have been made, and the population has expanded greatly. By 2007, there were more than 1,000 males, calling in 90 different ponds distributed over an area of about 100 km².

The next major project was to establish a treefrog population of mixed origin at mid-Sjælland. The site chosen for this was the military training ground at Antvorskov close to Slagelse. In a cooperation between West Sjælland County and the military, 30 ponds were made or restored there during the years 1993–1995. Treefrogs' eggs were collected at the two small populations in south Sjælland, at Stensby and at Præstø. During the years 1994–1998, a total of about 3,000 offspring were released at the new site. Results for the first two years indicated a survival of 7–8%, when the animals had been released as late tadpoles. The first 11 males were heard calling in 1996, and the population soon grew to about 500–1,000 males in the military area from 2002 onwards. In addition, the frogs have expanded through natural migration and have successfully colonized ponds up to 10 km away from the military area. Additional ponds have been made in the newly colonized areas to consolidate the expansion.

The theory to be tested by this project was that low survival of eggs from south Sjælland was due to inbreeding. Eggs were collected from the new population after about ten years (2005–2006) and compared with eggs collected from the two original source populations collected in the same years. The result is shown in the bottom row of Table 67.3. This mixed population had a higher rate of survival of eggs than did any other population in the table (the difference was significant). Unfortunately, there has also been an illegal transfer of treefrogs from Aarhus to Slagelse, and as the viability of eggs has always been high in the population used for release at Aarhus, this precludes drawing firm conclusions. We can conclude, however, that when we have a population with a high survival rate of eggs (i.e. high fitness), then this population shows a high potential for expanding and colonizing new ponds.

Other official projects aimed at establishing new populations are in the municipality of Fredericia south of Vejle (first release was in 2013) and at Roskilde on Sjælland (first release was in 2015).

There have been a number of illegal releases also. Treefrogs from the population near Præstø on Sjælland have been released at Knudshoved in southwestern Sjælland. The population there has grown to nearly 100 males and has expanded by several kilometres. Treefrogs from southern Jutland have been released on southwestern Fyn where they have established a stable population, and there have been a number of other illegal releases that have established small populations in some places.

In southern Jutland, treefrogs suddenly appeared west of Sønderborg, where they had not been before. They were probably taken from the nearby island of Als and released by a local forester. There is now a well-established population, and the frogs have spread to other recently created ponds further west.

E. Green toad, *Bufotes viridis variabilis*

Recent DNA analyses have revealed that green toads in the region of the Baltic Sea are a separate taxon, maybe even a cryptic species (Stöck *et al.* 2006). The epithet for this taxon should be *variabilis*. The mitochondrial DNA deviates greatly from that of *Bufotes viridis* in central Europe, so much that it seems likely that we are dealing with a separate species. However, as long as nuclear DNA has not been studied, the general opinion is that it is premature to speak of a separate species. So, until further notice, we shall call this taxon *Bufotes viridis variabilis*, with an expectation that later it may be elevated to full specific rank. This species/subspecies occurs in Greece, and from Turkey to parts of the Arabian Peninsula and the Caucasus region, and from there eastwards to southern Siberia. There are no known differences between *variabilis* and *viridis* in morphology, call or ecology. Danish green toads are remarkable in their tolerance of high salinity, and this might possibly be a specific feature of *B. variabilis variabilis*.

Green toads are limited to the southeastern parts of Denmark. The absolute border of the distribution to the northwest is the islands of Tunø and Samsø east of Jutland. The species is most widespread on the southeastern islands, eastwards to Bornholm.

Green toads breed in many very different types of bodies of water. The common feature is that the ponds have very little vegetation. Often they are temporary or newly dug ponds, for instance gravel pits. In some parts of the country, the main basis for maintenance of a population is flooding that arises when drainage pipes are temporarily blocked. Each flooding usually occurs over only one or two seasons. The offspring produced there migrate into the landscape. Green toads are regularly found in considerable numbers up to 3 km from the breeding site. When they become adult, new flooding may arise elsewhere, thus providing new sites for breeding. The frogs' loud calls serve to attract males and females from far away. As long as not all flooded areas dry out simultaneously, the toads may alternate their breeding between different ponds and in this way persist in an agricultural landscape that has no permanent ponds.

The green toad has been subject to severe decline. Out of the localities recorded in the 1940s, about 88% were lost when reinvestigated about 1980. This corresponds to a rate of decline of 45% per decade. This is gross decline, however, as the green toad is a pioneer species that often colonizes newly formed ponds and leaves old ponds that have filled in. So the net decline – which allows for colonization of new ponds – is not as great as the gross decline.

The net decline is known for regions where all occurrences, new as well as old, have been recorded. In the greater Copenhagen region, the net decline from 1983 to 1988, recalculated as decline per decade, was 59%. On the island of Samsø, the net decline during 1979–1988 was 68% per decade, and in the archipelago south of Fyn during 1980–1990 it was *ca.* 50% per decade. These figures for net decline are higher than the figures for gross decline up to 1980, which indicates that the rate of decline has accelerated severely. Obviously, by about 1990 it was clear that were nothing done, the species would be rapidly heading for extirpation.

On Samsø, the efforts to save the species started in 1986, when a pond was pumped dry to eradicate the fish. This caused a large breeding success there the year after. Up to 1996, a total of 35 pond projects were carried out, of which 29 were the digging of new ponds. Of these, 21 were colonized naturally by the toads, whereby the strongly declining population on the island got a boost and increased again. Some details are given by Amtkjær (1995).

In the archipelago south of Fyn, the species declined from occupying 22 localities in 1980 to being present in 11 localities in 1990. To stop the decline, pond projects were carried out, involving 34 ponds up to 1996. Twenty-three of these were new ponds. In 3 of these, reared offspring were released with success, and 11 others were colonized naturally.

Storstrøm County in southeastern Denmark, encompassing the southern islands and the southern part of Sjælland, is that part of Denmark where species are most widely distributed. Pond projects to benefit green toads began there in 1993. In the mid-1990s, the species occurred in at least 154 ponds. In about half of these, green toads had survived up to 2006, and in about one third of the ponds they had become extirpated. In the remainder, the situation varied from year to year. The net effect was a decline to at least 103 occupied ponds. However, without pond projects the decline would have been even greater. Up to 2005, a total of 107 ponds were made or restored. The result, as evaluated in 2006, was a clear progress for the species in 30 treated ponds, and extirpation in 11 treated ponds.

After 2006, when the county administration was abandoned, conservation of the species was managed by municipality administrations. Of special interest is a project in 2014 in the municipality encompassing Falster and east Lolland. There, advertising for interested landowners resulted in contacting more than 20 landowners. Many of them had green toads close to where they lived – under terrace flags, behind garbage bins, in greenhouses, in garages, in sheds or stables, in cellars or even inside the house. Often it was possible to document the presence of the species in places with no known breeding localities within many kilometres. Either the toads had already been seen before the first visit, or the landowner sent photographs afterwards as documentation. By making new ponds in such places, the presence of the species could be consolidated where former populations were not even known to have bred up to that time.

In the western part of Sjælland, nothing was done for this species until recently. Only few populations have survived. However, in recent years, a number of pond projects have been carried out, and the survival of the species has been secured even in places where it was not known to have bred a few years ago.

In most of northern Sjælland, the species had died out after 1980. However, the local stronghold of the species is close to Copenhagen, and there it has survived. A few ponds were made or restored about 1990, but most often the species finds and breeds in bodies of water that were not planned for it. It has colonized various basins dug for other purposes: temporary wet depressions in landfills, drainage ditches paved with flags or concrete, and even in basins in sewage plants. Migrations from old to new places seem to occur mainly when the toads swim in the sea along the coastline and in harbours. In addition, they also occur in more natural localities – grazed coastal meadows, especially on the island of Amager (part of greater Copenhagen) and on the island of Saltholm. Western Amager has many square kilometres of low-lying land behind the sea dike, where a large shallow lake was formed by a rise in water level. The small population of green toads barely surviving there began to breed in the newly flooded area, and hundreds of newly metamorphosed toadlets were seen emigrating from there over distances of many kilometres, in one case crossing a car and bicycle bridge to Sjælland. Overall, however, in the Copenhagen region, the situation is one of decline.

For many years, a large population of green toads has been known to occur on the low-lying grazed island of Saltholm between Copenhagen and Sweden. An EU LIFE project dealing with coastal meadows made it possible to improve the conditions there, mainly by adjusting the grazing pressure to the required level. Another interesting measure was to rebuild the basement of a farm building in such a way that toads could enter through holes into a large cavity under the house and hibernate there.

A large, new bridge between Denmark and Sweden was opened in 2000. In connection with the building of this bridge, a new island, called Peberholm, arose south of Saltholm. It was not intended to be a habitat for any threatened species, but it was soon colonized by green toads that most likely arrived at the new island by swimming from Saltholm or Amager. The animals have bred there with great success and, already a few years after 2000, the population had reached several thousand animals.

In the Baltic Sea, the green toad was formerly distributed over most of the island of Bornholm, often breeding in ponds at farms. It has declined severely and remains mainly in some clay and stone quarries, and in ponds close to the coast, including some rock pools. Only one pond has been made there specifically for the benefit of this species. One would expect that the toads would use some of the many ponds created for treefrogs, but this has not happened.

The green toad often occurs at places where large transport channels are being built – railway lines, roads, and especially bridges and ferry harbours. This has led to many compensatory measures to preserve the populations. The most marked such case was on Sprogø between Sjælland and Fyn, which is crossed by a motorway that opened in 1998 as a part of one of the world's longest bridges, the Storebælt bridge. The main effort was to construct a permanent fence in order to prevent the toads from crossing the motorway. However, the toads could swim along the coast and colonize the new landfilled areas on the other side of the motorway. They have formed large breeding populations there now. The effects of the measures taken – the fence, restorations of ponds, improved grazing – are monitored by regular estimates of the total population size by capture-recapture of marked individuals. The population remains fairly stable at about 2,000 animals.

In general, it is easy to help the green toad on a short-term basis, because it effectively finds and breeds in newly dug ponds. When ponds have been dug, for instance, for fire-bellied toads, they are regularly colonized by green toads whenever they live in the vicinity. However, it is difficult to have the ponds function on a longer term, because as soon as *Typha* or *Phragmites* start growing in the ponds, the toads stop breeding there. The best way to prevent this is to have the ponds grazed by cattle or other domestic animals. This is clear from evidence from the pond projects in Storstrøm county; there the species has survived in about half of the former breeding ponds that were still being grazed after restoration, but it disappeared from practically all the restored ponds that were not grazed subsequently.

Many of the pond projects for green toads have been carried out on coastal meadows and on small islands. There the ponds are vulnerable to flooding with salt water during winter storms. Many breeding ponds are too salty in some of the years, and only function as breeding sites after winters with no flooding by salt water. A storm hit Denmark in December 2013 and, in the following season, many former breeding ponds had high salinities. However, the ability to breed in such flooded ponds is remarkable. In cooperation with a German field station on the island of Fehmarn, we have studied the survival of newly spawned strings of eggs gathered on Lolland in Denmark, and on Fehmarn. Eggs from both populations survived and developed through metamorphosis when placed in tanks with salinities up to seven parts per thousand. At eight parts per thousand, very few survived, and at nine and above, none survived. Under field conditions, however, conditions may be more favourable. When water gradually evaporates from the ponds in May and June, the salinity increases, so high salinities are reached only when tadpoles are already well developed. In many cases in 2014, live, healthy tadpoles were found at salinities up to 10 parts per thousand, and live, but only slowly growing tadpoles were found at 11 parts per thousand. These salinities are just as high as in the surrounding sea. In such cases, the tadpoles may be found – sometimes in huge numbers – swimming with shrimps and even sand gobies. Although green toads may occasionally call from seawater, they rarely, if ever, breed there, presumably because of the (larger) fish.

In cases in which populations of green toads have been boosted by pond projects, for some years there may be large numbers of toads in the landscape. In such cases, we observe colonization of other islands or peninsulas in the vicinity. Adult toads occasionally are seen swimming in the sea, and there is little doubt that emigrating toads may cross stretches of sea of several kilometres to colonize other islands or parts of the mainland.

In places that cannot be reached by toads in natural ways, an option for us is to breed offspring and release them at new sites. Such projects have been performed in a number of cases. Sometimes the situation is simple – eggs are spawned, e.g. in flooded fields that are expected to dry out. So if such eggs are found, they are taken and reared until the tadpoles have hatched. These are then transferred to some suitable pond, e.g. if a pond has recently been dredged in the vicinity. In a few cases, new populations have been established in this way.

In other cases, more elaborate breeding projects have been carried out. An example is a site on the island of Falster, where a pond was dredged in the autumn, but in the next spring, no male green toads remained. All that was left was a single female that sat waiting at the water's edge night after night. Then a large aquarium was prepared. The female was caught in the night and transported in a bucket to another locality 15 km away, where a few males were calling. A male was put into the bucket, and it entered amplexus almost at once. The couple were then transported to the aquarium. The male left its position on the female, but soon entered amplexus again. The couple remained in amplexus in the aquarium for three days, until spawning finally occurred. The female – old and large – produced about 19,000 eggs, thereby confirming statements in the literature that green toads may produce up to *ca.* 20,000 eggs in a single mating.

Parts of the strings of eggs were reared, and the resulting newly metamorphosed toadlets released at suitable sites. Sometimes such rearing-and-release projects have been successful, but have also often failed. It seems that suitable terrestrial habitats are crucial for the survival of the released toadlets.

F. Natterjack toad, *Epidalea calamita*

The natterjack toad is generally distributed throughout Denmark. However, it generally becomes more common as one goes from east to west. From the 1940s until *ca.* 1980 it disappeared from about 65% of its former localities. Since then, the rate of decline has increased, especially in eastern Denmark. In the greater Copenhagen region, the net rate of decline during the 1980s was 17% per year, corresponding to 84% per decade. This is a steeper decline than for any other amphibian species. On the rest of Sjælland and the other eastern islands, it has also declined severely up to about 1990, and it is now definitely rare on the islands east of the Great Belt. This is also the case on Bornholm, except for some coastal localities with stable populations, especially in rock pools. In the period 1980 to 1990 it disappeared from 13 out of 21 regions on Fyn, a decline of 62%. During the same period, it disappeared from two out of 10 islands and islets in the archipelago south of Fyn.

The rate of decline in Jutland is not known with a similar precision, but we have counted the number of larger squares (10 km x 10 km) in which it occurs. From the period 1976–1986 to the period 2002–2012 it had disappeared from about 10% of the squares along the west coast of Jutland, and from more than 20% of the squares around the large fjord Limfjorden in north Jutland – which is the area in Denmark where the populations are most dense.

The present population of adult animals is estimated to number about 10,000 individuals along the western coast of Jutland, 5,000 around Limfjorden, and 10,000 in the rest of Denmark. Obviously, the situation in about 1990 was the same as for other threatened amphibian species. Immediate action was necessary. However, more focus was on other species than on the natterjack toad and not many pond projects were carried out during the 1990s. In the Copenhagen region about 15 ponds were made, and there was a temporary recovery of natterjack populations on the island of Amager, close to Copenhagen. However, in the long term, the species has not survived in that region, neither on Amager nor elsewhere. One of the problems has been that additional ponds made about 2000 were not prepared by experts with sufficient experience, and the ponds were dug too deep.

On Bornholm, three ponds for the natterjack toad were made at one locality. However, there too they were dug too deep, and they only favoured amphibians other than the natterjack.

In Storstrøm county, the first pond projects for the natterjack were carried out in 1992 and 1993. This includes a site with only one, maybe two, calling males from the year before, and the nearest other site with about 5 males, 7 km from the first. In both these places, the natterjack toad population has increased and actually persisted up to at least 2015, apparently with rather stable populations. Migrating individuals are found regularly midway between the two localities, so there must be some genetic exchange between them. There has been a case of spontaneous successful colonization farther along the coast, 17 km from the nearest of these two localities.

A similar effort on south Sjælland failed. There are other localities in that region, however, where the natterjack toad survives, in some cases in large populations. These are mutually isolated natural areas and nature reserves. However, increasing salinity in the bodies of water is a problem at some of these sites, and may be a severe threat over a longer term.

There is an area with many large gravel pits at a single inland locality in central Sjælland. Some of these pits were preserved especially for the toads a few years ago, and the toads are still there, but the pits have filled in with reeds and breeding is no longer successful there. However, the toads do successfully breed in a few other active gravel pits nearby. Still, this isolated population is severely threatened by inbreeding.

Farther west, the county of Fyn made about 15 ponds for the natterjack during the 1990s, especially in the archipelago south of Fyn, and this did halt the trend towards extirpation at a number of places. Three additional ponds were made in 2004, and 77 more during the years 2012–2015, mostly on the main island, but also on some small islands in the strait between Fyn and Jutland. Breeding success has been reported in several of these new ponds. Of special interest is the project at Bøjden on southwestern Fyn at an estuarine reserve for birds. Additional land was purchased around the estuary, six ponds were dug there, and a grazing regime suitable for natterjack toads was established. Toads colonized all of the ponds and breed in some of them.

In east Jutland, the natterjack toad was formerly widespread, but it is now rare and still declining. In recent years, more than 40 ponds have been dug or made deeper in the former Vejle county, but with limited success. On one island close to Jutland (Hjarnø), the efforts came too late, and the population has now perished. In four other localities, the efforts were in time, and the population has survived. However, all these populations are now so small and inbred that they will probably die out unless further measures are taken, e.g. artificial breeding and cross-release between sites.

In several places in southern Jutland, some pond projects have recently been carried out in gravel pits in the inland, paid partially by the municipality, and partially by private funds. These projects consisted of scraping very shallow depressions, which subsequently were successfully colonized by breeding toads. It is uncertain whether this will continue, however, as no subsequent management has been planned.

On the island of Als close to southeastern Jutland, the natterjack toad has survived, but barely so. For many years, nothing was done to preserve it but, from 2012 onwards, a few depressions have been scraped, and grazing has been improved, financed mainly by the municipality.

On the island of Mandø in the Waddensea, the natterjack toad population had become very small, so the county paid to have some depressions scraped in 2005. When the site was monitored in 2013, there was still successful breeding in some of these depressions, some of which, but not all, were grazed. Thus, the effect of these efforts lasted for at least eight years. Apart from Mandø, nothing had been done to improve the situation for the natterjack toad along the western coast of Jutland for many years. In recent years, however, a number of EU LIFE projects have been carried out; these focused mainly on general improvements of heathland and dune habitats, but did include measures of benefit to the natterjack toad, such as improved grazing and restoration of natural hydrology, e.g. by blocking ditches. The localities that were involved stretch from the south (island of Rømø) to near Skagen (the northern tip of Jutland). The first of these LIFE projects ran from 2002 to 2006; there were a few cases of natterjack toads calling up to half a kilometre from the nearest known places where the species was reported 5–7 years previously.

Natterjack toads are numerous in the military training area near Blåvandshuk (the westernmost point in Jutland). They breed mainly in the tracks created by tanks, but reduced military activity has resulted in fewer tracks and fewer breeding sites. Some new ponds were dug there during the LIFE project, and natterjack toads started calling up to one kilometre from previously known sites. North of this place a few depressions were scraped, and this too was successful as afterwards the toads bred in those depressions. Only a few kilometres from this site lies the recently restored, large, shallow Lake Fiilsø; this lake potentially provides the natterjack toad with suitable breeding sites, but had not been colonized up to 2015.

Another large LIFE project encompassing 51 wetland sites with a total area of 529 ha was initiated in 2014. It is still too early for results to be known. Another new project dealing with dune habitats and covering 1,137 ha will begin in 2016.

Other recent EU LIFE projects deal with coastal meadows near Limfjorden. The main elements are the scraping of shallow depressions and the fencing of areas to facilitate grazing. The first six depressions were scraped in 2015. At several sites, formerly large populations have been reduced to 10–20 calling males; it is expected that some of these populations will again increase owing to grazing of the wet depressions.

There used to be many shallow, grazed ponds, meadows and flooded fields for natterjack toads in inland localities in the northern half of Jutland; nothing has been done to benefit these populations up to the present time and, in recent decades, some of them may already have disappeared.

G. Mixed populations of water frogs

The edible frog, *Pelophylax* kl. *esculentus*, is not a species in the usual sense of the word. It originated through hybridization between the pool frog, *Pelophylax lessonae*, and the marsh frog, *Pelophylax ridibundus*. The hybrids are most unusual in that the parent species' genomes remain unchanged, without any interchange of genes (there is no crossing-over during meiosis). Therefore, a hybrid frog has one intact *lessonae* genome, designated as L, and one intact *ridibundus* genome, designated as R. Thus, a diploid hybrid frog has two genomes, one L and one R and is termed an LR frog. There are also triploid frogs, some of which are LLR, and some of which are LRR. The LLR frogs are similar in their morphology and in their calls to *lessonae* frogs (which, of course, are LL), and LRR frogs are somewhat similar in morphology and calls to *ridibundus* frogs (which, of course, are RR). All frogs that have both L and R genomes are called *esculentus*. Populations of *esculentus* often contain all three possible forms, i.e. LLR, LR, and LRR.

In most of Denmark and in neighbouring regions (southern Sweden, northernmost Germany), *esculentus* exists in pure populations without either parent species (Christiansen *et al.* 2005). These populations always have at least two genotypes, i.e. LR and LLR, or LR and LRR. On the island of Bornholm, the situation is different, because RR frogs, i.e. true marsh frogs live there in addition to LR, LLR and LRR frogs (Rybacki 1994). These marsh frogs are, however, different from those in central Europe in some respects. For instance, they often breed in small ponds, whereas in the rest of Europe, marsh frogs normally live in larger lakes and along rivers.

There are indications that the marsh frogs on Bornholm originally colonized Bornholm about 9,000 years ago when the island was the tip of a peninsula connected to the site where the Oder river now flows into the Baltic Sea. The water frogs that live in the Oder river today are unusual in that marsh frogs, of both sexes, live together with *esculentus* frogs that are represented only by males. So the male *esculentus* (LR) mate with female *ridibundus* (RR), and the resulting offspring is again male LR and female RR. In principle, the RR males are not necessary, but, being present, they mate with females of their own species, and produce offspring that is again male and female RR.

At the mouth of the Oder river is the island of Wolin. There, triploid frogs occur in the populations. These are males and females of the LRR type. They mate with the other types of frogs and produce eggs and sperm that are R, but mostly with female factor, whereby most RR offspring will be females. So here, *ridibundus* are mostly females.

Now, on the eastern part of Bornholm (but not on the western part), there are rather similar populations; in most ponds on east Bornholm, RR males are rare or absent. That is, marsh frogs do occur, but mostly as females, i.e. not as a pure species that can breed by itself. These females mate with some type of *esculentus* males. In some ponds, evolution has continued to the point where all males are LR and all females are RR. Animals from such populations have been studied by crossing experiments in captivity; again the offspring are exclusively LR males and RR females. The males have 'adapted' to this unusual situation by calling, not as normal LR males, but as RR males, thereby attracting RR females. Such populations are most unusual.

As far as is known, nothing like this exists anywhere else in the world. They are a unique example of evolution that has proceeded since marsh frogs entered Bornholm about 9,000 years ago, and thus are certainly worthy of protection. Strangely, the type of population may vary from pond to pond over distances of less than one km. For instance, there are populations that are nearly exclusively RR males, living close to those with no RR males, but they do not mix. Altogether, the variation on Bornholm is enormous, with about ten different types of populations of water frogs represented on the island.

These complicated facts were clarified by investigations made in the 1990s, carried out by a Polish specialist in water frogs (Rybacki 1994). Afterwards, a few new ponds were made and others dredged to support these peculiar populations. Those ponds were colonized immediately by frogs from populations nearby.

A very peculiar situation exists on the small island of Christiansø, 20 km to the north of Bornholm. This island and the neighbouring Frederiksø Island consist exclusively of rocks. There are a number of ponds in depressions among the rocks. Frogs from Bornholm were released there on private initiative in about 1950. They have lived there since, were once very numerous, but have since declined severely. It was unknown whether they were edible frogs or marsh frogs until 1997, when they were investigated by Fog (1997) and Rybacki (1999). Nearly all males were LR, and nearly all females were RR. The males' call is somewhat like *ridibundus*, although not completely so. This incongruence between their call and their morphology contributed to the confusion as to their true identity.

This population declined owing to release of various fish in their ponds, especially eels and crucian carp. The fish were predators on the tadpoles and had become so numerous that they prevented any breeding success by the frogs. The situation was further aggravated by the very uneven sex ratio – the number of male frogs was about eight times higher than the number of female frogs. In 1997, only four females remained, and some of these were infertile. In the autumn of 1997, inhabitants on the island tried to eradicate the fish in the most important breeding pond, but were not entirely successful. In the next year, 1998, only two females remained. One was old and severely wounded, presumably due to numerous attacks from the vast surplus of males. So, presumably to avoid attacks from the males, the females kept away from the breeding pond and did not breed that year.

New plans were made to dredge two ponds in the autumn of 1998. This could not be carried out by digging machines because of the rugged rocky bottom. Instead, an entirely different method was applied. Large tubes were laid across the island, and seawater pumped from the harbour into the pond to flush out all sediment. Another pump in the pond pumped the water out again and over the other part of the island back into the sea. By this method, ponds could be cleaned completely, and fish of all species removed. This was done in the autumn and by the spring of 1999 the ponds contained fresh rainwater again.

In June 1999, the island was visited again. A female marsh frog, which had been kept in captivity, was placed in a breeding cage in the main breeding pond, and was injected with hormone. However, she remained infertile. The only other surviving female was found in amplexus with a male in a very small, shallow pool next to the pond. The couple was caught and placed in the cage, where they remained for two days. Another male jumped into the cage and also went into amplexus with the female, so when she finally spawned her eggs after two days, they presumably had been fertilized by both males. The eggs were collected. Some were put in an aquarium and a local school teacher looked after them. The other eggs remained in the cage and grew as tadpoles there until they were set free two months later.

In the autumn, newly metamorphosed froglets, not only *esculentus*, but also *ridibundus*, were seen at the edge of the pond. In the following year, young frogs of both sexes were sampled for DNA analyses.

In 2008 the island was visited again. Now the total population had grown to 220 individuals, of which 69 were adult females and 21 were juvenile females. They were distributed among 11 ponds on the island – of which several more had now been dredged – and a few individuals were found in a pond on the neighbouring island of Frederiksø, which they could easily reach by swimming in the sea.

Thus, this remarkable population was saved just in time. It has considerable touristic and scientific value. All frogs descend from one single mother but, because of the peculiar mating system, inbreeding depression cannot arise in the LR males in this type of population.

VII. References

Amtkjær, J., 1995. Increasing populations of the Green Toad (*Bufo viridis*) due to a pond project on the island of Samsø. *Memoranda societatis pro fauna et flora Fennica* **71**: 77–81.

Andersen, L.W., Fog K. and Damgaard, C., 2004. Habitat fragmentation causes bottlenecks and inbreeding in the European tree frog (*Hyla arborea*). *Proceedings of the Royal Society of London B* **271**: 1293–1302.

Bisgaard, A., Knutz, H., Mikkelsen, K., Mikkelsen, U.S. and Weitemeyer, L., 1979. Bjergsalamanderen (*Triturus alpestris*) i Danmark. Nyere undersøgelser i 1975–78. *Flora og Fauna* **85**: 27–36.

Bringsøe, H. and Mikkelsen, U.S., 1997. Newt in progress: Status for *Triturus alpestris* in Denmark. *Memoranda Societatis pro Fauna et Flora Fennica* **73**: 105–8.

Christiansen, D.G., K. Fog, B., Pedersen, V. and Boomsma, J.J., 2005. Reproduction and hybrid load in all-hybrid populations of *Rana esculenta* water frogs in Denmark. *Evolution* **59**: 1348–61.

Collin, J., 1870. Danmarks frøer og tudser. Naturhistorisk Tidsskrift 3. række 6: 267–352.

Fog, K., 1988. Reinvestigation of 1300 amphibian localities recorded in the 1940s. *Memoranda Societatis pro Fauna et Flora Fennica* **64**: 134–5.

Fog, K., 1993. Oplæg til forvaltningsplan for Danmarks padder og krybdyr. *Miljøministeriet, Skov- og Naturstyrelsen* [contains English summary].

Fog, K., 1997. A survey of the results of pond projects for rare amphibians in Denmark. *Memoranda Societatis pro Fauna et Flora Fennica* **73**: 91–100.

Fog K. and Andersen L.W., 2005. DNA-analyser og naturforvaltning – løvfrøer som eksempel. *Naturens Verden* **2005**: 2–12.

Fog, K. and Wederkinch, E., 2016. Does legal site protection lead to improved conservation of ponds with fire-bellied toads *Bombina bombina* in Denmark? *Conservation Evidence* **13**: 18–20.

Hansen, F., 2004. Verbreitung und Gefährdung des Laubfrosches (*Hyla arborea*) auf Bornholm (Dänemark) und Massnahmen zur Lebensraumoptimierung. Ed. D. Glandt and A. Kronshage. *Der Europäische Laubfrosch. Zeitschrift für Feldherpetologie, Supplement* **5**: 133–43.

Hanström, B., 1927. Marint massuppträdande av Bufo viridis i Öresund. *Fauna och Flora* **22**: 249–57.

Hvid, M., 2005. Holm-Andersen strimlerne. *Natur og Miljø* **2005, no. 2**: 9–10.

Lütken, C.F., ca. 1847: Handwritten manuscript presenting the lectures given by professor J. Steenstrup 1846–1848. Kept at the library of Zoological Museum of Copenhagen.

Pfaff, J.R., 1943: De danske padders og krybdyrs udbredelse. *Flora og Fauna* **37**: 49–123.

Rybacki, M., 1994: Water frogs (*Rana esculenta* complex) on the Bornholm island, Denmark. *Zoologica Poloniae* **39**: 331–44.

Rybacki, M., 1999: Grøn frø bestande med Rana ridibunda hunner og Rana esculenta hanner. Pp. 109–121 in P. Wind *et al.*: *Overvågning af rødlistede arter 1998*. Naturovervågning. Arbejdsrapport fra Danmarks Miljøundersøgelser nr. 110.

Skriver, P., 1988. A pond restoration project and a tree-frog (*Hyla arborea*) project in the municipality of Aarhus. *Memoranda Societatis pro Fauna et Flora Fennica* **64**: 146–7.

Skriver, P., 2007. Overvågning af løvfrø i Århus kommune. 2007. *Rapport udarbejdet af Aqua Consult for Århus Kommune, Natur og Miljø*.

Stöck, M. *et al.*, 2006. Evolution of mitochondrial relationships and biogeography of Palearctic green toads (*Bufo viridis* subgroup) with insights in their genomic plasticity. *Molecular Phylogenetics and Evolution* **41**: 663–89.

Index

Page numbers in *italics* indicate figures or tables. Please note that listings are under species name, not under family.

CPSIA information can be obtained
at www.ICGtesting.com
Printed in the USA
BVHW010437100719
553032BV00002B/4/P

9 781784 270162